by NAH*R*
Nature, Art & Habitat

Transect of Coexistence

Inquiry into Nature, Art, and Habitat

chief editor
Ilaria Mazzoleni

editors
Maristella Casciato
Lucinda Coleman
Sara Invernizzi
Roxanne Korda
Richard Molina
Gabi Scardi
Aslı Suner
Deborah Weintraub

To the Valleys and Mountains,
to all Species that journey with us,
to our Ancestors that moved here and made it home,
to the late Alessandro Mendini for his presence always,
and to our Families.

Contents

Foreword

Malena Martinez

Transect of Coexistence: Inquiry into Nature, Art, and Habitat invites all the curious and inquiring inhabitants of this richly biodiverse planet to seek opportunities to exchange diverse ideas; to consider humanity's impact on ecologies and embrace diversity. To travel this path can be wonderfully rewarding and exhilarating, journeying down roads we could never previously have imagined or even dreamed of. There is tremendous pleasure in expanding networks of peoples and efforts, and in re-thinking the ways of our economic, agricultural, social systems that now need to evolve to meet our expectations and the needs of the planet. This book presents the power of interconnection, and the beauty of such organically found encounters. Consider how when a piece of fabric is handmade, it is the knowledge and participation of many elements including ancestral wisdom, the arts, and craftsmanship which are the threads required to weave the final product. This art of constructing a unique textile from fibers that are joined and arranged in a particular way is magic! It is the magic of bringing people together and sustaining a conversation about the present, and human ways to find truths and wellbeing through stories. The various techniques human groups use to develop textiles, including using various resources like vegetable fibers as threads, can depend on the place of origin. As such, the result is an integration of meaningful cultural and symbolic patterns of special design alongside a collaboration with nature. Some patterns are very complex, others less so: more literal, and almost always related to the environment (such as the representation of mountains, rivers, birds) or with a very personal story.

Let me share with you my story of a winter afternoon in Lima. My brother Virgilio, and I, were just beginning to imagine the likelihood of making something more our own which responded better to who we are. We decided to take advantage of the time and free tables after the lunch service to receive the Peruvian geographer and professor Edwin Gabriel Campos. With some suspicion, he moved the chair to accommodate himself, seeking to understand from our gestures what we expected to hear from him. "Very simply," Virgilio hastened to clarify, "we want to know what makes Peru the biodiverse place they say it is". At that moment, his obvious enthusiasm made him sit up and take a sheet of paper from a notebook he had, to graph the first, and most powerful of the reasons. "See," he continued, "the territory of Peru is like a crumpled piece of paper". He took the graph paper and squeezed it tightly between his hands. Spreading it out slightly, he explained that the folds resembled the various different life systems of a mountain range that cross the country from north to south. Then also there were the other relevant geographic components such as the preponderant presence of the Amazon (more than 60% of the territory), the two opposing temperature currents of a Pacific sea of special density due to the geo-location, and the trade winds from east to west. More and more aspects were added to understand what could have at first seemed a simple question to answer but in truth required a detailed knowledge of the many coexisting living and natural systems across Peru.

Overtime, other conversations including with anthropologists, historians, artists, artisans, curious journalists, and many local people in rural areas led us to think about this crumpled paper and the intricacies of inhabiting an immensely complex territory. The geographical heterogeneity is accompanied by an important ethnic diversity. And these populations developed mastery in overcoming the challenges of this irregularity of systems: the extreme altitude, at more than 4,000 meters, the flooding of mighty Amazonian rivers, the swings of the dark blue sea. They learned to design their existence based on marked seasons, or simply wait for the arrival of the rain and trust that the gods would protect, provide what is necessary, and also expect retribution. These same people who seemingly tamed nature's surroundings created a relationship forged in respect, preserving the natural balance in harmony with the world of which they know they are not the center but are merely a component. I find it fascinating when villagers talk about the effects of climate change as something tangible; it indicates the closer one's connection to surroundings, the more open your eyes will be to reality. There are so many lessons to learn, for us urban citizens.

What I am describing is the story of the creation of our interdisciplinary center *Mater Iniciativa*, today based in Lima and Cusco, Peru. It was born from the heart and hands of two siblings who at first recognized ourselves as ignorant, then began to search for ways to articulate the world whilst promoting the growth of this incredibly ecologically diverse country. Virgilio, the curious explorer chef, whose passion exceeds the limits of the infrastructure of a restaurant (or several), and I, a medical doctor by profession, who found my vocation in the opportunities with food and its context of origin or existence, to expose its impact on overall health.

Mater's mission, alongside the restaurants in Peru, is to work as a laboratory for participatory culinary proposals and explorations that lead to collective growth. We see gastronomy as a vehicle capable both of boosting the economy of our entire country, and making visible our cultural, ecological, historical, geographical, and narrative richness. It turns out that humans fundamentally feed themselves, and whether it is with food or information, we are constantly seeking to be nourished with more, and if we are really conscious, of better quality. As you taste a recipe the food in your mouth is also a message to the mind. So, when visitors experience *Central*, or *Kjolle* in Lima, or *Mil* in Cusco, they will go on a journey, transecting distant Peruvian territories and cultures in ingredients, and learning how the origin of the different ecosystems is determined by factors such as the altitude at which they are found.

Just as this book follows an interdisciplinary path, bringing together knowledge gained through scientific study, deep observation, and artistic practices, through *Mater* we have also designed and implemented scientific, as well as artistic and cultural projects. We distinguish these two great areas, yet they often converge, because at the end of the day there is always human connection within each step of these processes. Great examples are the beverages fermented according to ancestral Andean techniques, which today we can mimic step-by-step in controlled systems regarding microorganisms and by-products. And then the cultural projects related to the art of ceramics, painting, textiles, and in general, the design and development of utilitarian objects that then participate in the offered gastronomic experiences. On our tables, pieces of wood from rescued trunks from the Amazon meet with pieces of goldsmithing with Peruvian metals.

Our experiences are created very much like the molding of ceramic pieces, with imperfections arising from the pressure marks of the fingers of those who work on them. Similarly as you read this book the imprints of others are as fascinating and full of history as the Amazonic looms, and the strokes of brushes embedded in clay, that keep memories of peoples and the cosmos. The commensal dialogues between peoples who speak different languages (whether literally or through different disciplines) allow for a common and globally appreciable language to flow around ingredients that until that conversations were unknown: this too you will discover as you read the pages to follow. Consider your reading of this book a discovery of ingredients, just like that day my brother and I sat at that table. Every page, similar to every meal, should mark a 'before' and an 'after'; that is how we know it is meaningful, and it has the potential to grow and become something more, or stay exactly as it is, with the understanding it is worth preserving and it should not change.

This book is a symphony of individual melodies, and creates music that celebrates experiences of the past and present, as views of worlds that are expressed through personal stories, expert knowledge and artistic practices.

For us empathic readers and researchers at heart, a myriad of views make our visual field widen. This curiosity is important to understand what occurs, and to propose a way to build not from a calamity and crisis perspective, but from the intention of solving in community and the complete certainty that knowledge is the foundation of better decision-making and the co-creation of valuable opportunities for more.

Let us seek these encounters to continue thinking about ourselves and enjoy the windows that each story offers us to look into worlds, so that in celebrating biodiversity we will be celebrating humanity and our part of the cycles of life and nature.

Preface

NAHR Editors

Current ecological crises are provocations that call for actions. Our lived experiences help us to react insightfully, making best use of acquired knowledge from different disciplines, cultures, and epistemologies. In this book, we seek to share our stories of response to fundamental questions concerning nature and culture, contributing to contemporary discourse regarding the climate emergency, and embracing a polyphony of voices - a key principle reflected in the goals of *Nature, Art & Habitat* (NAHR), *an eco-laboratory of multidisciplinary practice* that has served as both catalyst and producer of this publication.

NAH*R* felt an urgency to gather people from diverse backgrounds and facilitate environmentally conscious approaches to both the making of creative actions and deepening of ideas exchanges across our perceived borders. In gathering, there is a sharing of perspectives, within and beyond the village community. Fellows offer their thoughts, develop artifacts, and engage meaningfully with others. These actions remain essential to the participatory aspects of NAH*R* and are reflected in the contributions to this book.

1 - Remi Picó. *Random 0*, 2023. (Courtesy of the artist)

NAH*R* was established in 2016 with an antispeciesist and eco-centric vision by Ilaria Mazzoleni, in collaboration with Anna Santi, Alexandru Balasescu, Alberto Mazzoleni, and Alessandro Mendini, by whom we will always feel inspired. NAH*R* is mainly dedicated to hosting residencies in the rural Taleggio Valley, Italy—the birthplace of Ilaria Mazzoleni—with an outpost in California since 2021. Since the start, it has also been supported by the Ecomuseum Val Taleggio, a regional social entity operating since 2008 in the Valley.

The residencies offer a lively space to reflect and create, collaborate and interact, and take full advantage of the surrounding landscape, beyond any academic or institutional restrictions or requirements. NAH*R* has become a place for immersive experiences which animate a dynamic and multidisciplinary think-tank for field work, inquiry, and ideation.

With the book now in your hands, NAH*R* hopes to share the achievements and reflections triggered through about a decade of inquiries into nature and culture. From our point of view, these achievements and reflections represent a concrete call for action.

We believe in slowing down, observing, inquiring, seeking out the invisible, and—perhaps most importantly—traversing the many disciplinary divisions of today to find out what we have lost during modern times. Dealing with the contemporary world requires sincere attention: engaging the "arts of noticing", as expressed by anthropologist Anna L. Tsing, and abandoning the specializations of the twentieth century to give way to a twenty-first century call for interconnectedness and interplay.

This book is designed to be as inclusive as possible whilst also premised on the ethics of environmental awareness. It is a collection of essays that includes written and visual materials from various specialities. At this moment of global climate emergency, our offering of entangled texts hopes to suggest future pathways for healing the relationship between the built and natural worlds. Some authors are invited writers, some are friends of NAH*R*; others are former NAH*R* Fellows whose work is shared with generous perspicacity.

Start reading anywhere. Explore the text thematically or sequentially, approach with curiosity in your own way. Begin with the artist or scientist, the anthropologist or the environmentalist. Each chapter is delicately integrated, shaped as a proactive polyphony, building hope against the current ecological crisis.

Journeying NAH*R*'s Transect

Ilaria Mazzoleni

A Climate Emergency is upon us. We, all living organisms on Earth, are in one way or another facing this challenge in our daily lives. As species adapt or go extinct, humans struggle in trying to find meaningful ways to mitigate, rethink, respond, and to change our current models designed for inhabiting and operating on this planet, towards a new paradigm conducive to all life. *Transect of Coexistence: Inquiry into Nature, Art, and Habitat* joins the discourse by seeking new insights into our environment, suggesting ways to combat the ecosystem collapse that can lead to change, and offers ideas to think differently and proactively about the types of impact our lives can have, from individual gestures to collective actions. There are many activists, environmentalists, and artists who are working to address these issues; within the pages to follow you will discover even more!

The inspiration behind this work lies within the debate concerning the fundamental meanings of the terms *nature* and *culture*, and their rhizomatic[1] connections. We will come to articulate later why we have chosen the term *nature* instead of ecology, even though our practice is mostly premised on observing *nature* from a scientific starting point. Polymath, Francis Galton, coined the phrase 'Nature versus Nurture' in the mid-1800's. We embrace an evolved version of this phrase, 'Nature and Culture', replacing nurture with culture, and the opposition of *versus* with the conjunction *and*. In doing so, we acknowledge how deeply interconnected we—our human-based perspectives and endeavors—are with the natural world, however strongly we desire to de-center our viewpoint[2]. Many of our reflections and work move along this *fil rouge*, pondering, thinking, and acting upon the desire of unfolding their layered and trans-sectional articulations to ultimately reestablish the balance we are striving for as we continue to share planet Earth. Perhaps the most challenging aspect is found in moving away from established and destructive binaries, as explored by Rosi Braidotti (2017) through the lens of the posthuman, recognizing complexity as key factor, and acknowledging then working with the coexisting multilayered interrelations of all elements. In many ways, our dominant culture has traveled further away from nature,

[1] Rhizomatic refers to a complex lattice of networked and non-hierarchical connections, as described by Deleuze and Guattari in *A Thousand Plateaus* (1993).
[2] NAH*R*'s conceptual grounding in its appreciation of multispecies histories and the nature and culture divide is also rooted in Donna J. Haraway's term "naturecultures" (2003). It is recommended to explore this concept and Haraway's seminal work starting from *The Companion Species Manifesto* (2003).

but as nature is essential to who we are, NAHR prioritizes the positive foregrounding of it, to regain an understanding of our entanglements, and to heal from the ongoing environmental crises within a post-anthropocentric vision.

Consequently, this book aims to interrogate and present the multi-directional learning which we call a transect. An immersive, participatory, and active observation in the field, along the transect, is in fact a deep experience of transferal which can be translated into exploratory, scientific, and proactive actions that can stimulate inquiry-based contributions.

In our increasingly globalized 'online' lives, such fieldwork is acquiring further critical value. By extricating ourselves from the fast track of contemporary urban life, we can slow down, dissect issues accurately, and we can find deeper and more meaningful reconnections to nature. Fieldwork has provided key knowledge systems throughout time, together of course with empirical studies.

As a scientific tool, a transect provides a line or section along which to take observations and measurements of environmental factors. Along the transect we are exploring, we take a broader view of the global situation and spend time noticing details of the local context: the exploration is of a physical ecological condition or of a social environmental one, or a combination of both. In this way, creative and scholarly actions engage with the world by shining a light on single moments, objects, or thoughts, whilst thinking laterally about the relations between nature and culture, urban and rural, and connecting the local with the global.

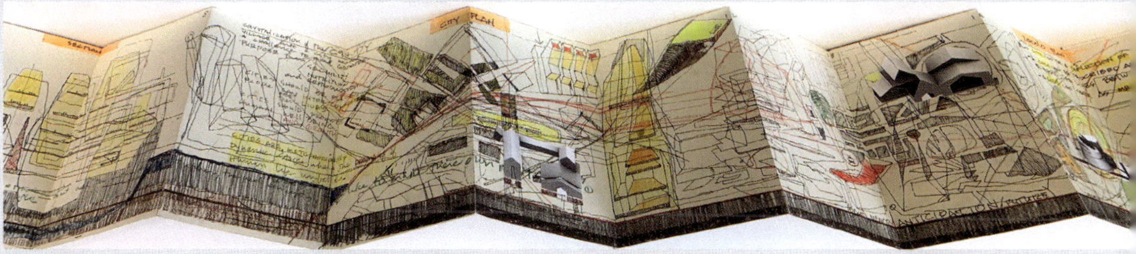

1 - A Transect: a study and speculation in section, plan, and massing, of human settlement. Michael Rotondi. *J-fold*, 2020 (started) and 2023 (resumed and finished). (Image courtesy of architect)

EMBRACING INTERCONNECTIONS

Working with models taken from ecology urges a view of the system versus the individual, and moves our center to a position among "the other" promoting paths towards empathy, and actions from the eco-centric, rather than ego-centric, perspective. The rural—which includes everything outside urban concentrations—is sometimes seen as a periphery, despite its fundamental role in urban life through its ecological richness. If we recognize this dependency and learn to foreground nature in our decision-making, then what was once peripheral can take center stage, as its biodiversity is the essential basis for all lives. Growth driven economies have made us design and prioritize extractive modes of operation which move away from being attuned to Earth-conscious thinking and actions. In giving more attention to nature we can learn to redesign our cities to work with, and not against, the rural. Getting closer to understanding the interconnected environment might help overcome biophobias which promote separation and greatly limit the possibilities of coexistence and the nurturing of biophilic[3] lifestyles. Accepting a mindset that is inclusive of urban and rural as complementary parts of a larger system and beginning to change the way we inhabit this planet could provide a significant decrease in humans' footprint, and be critical to the survival of all species.

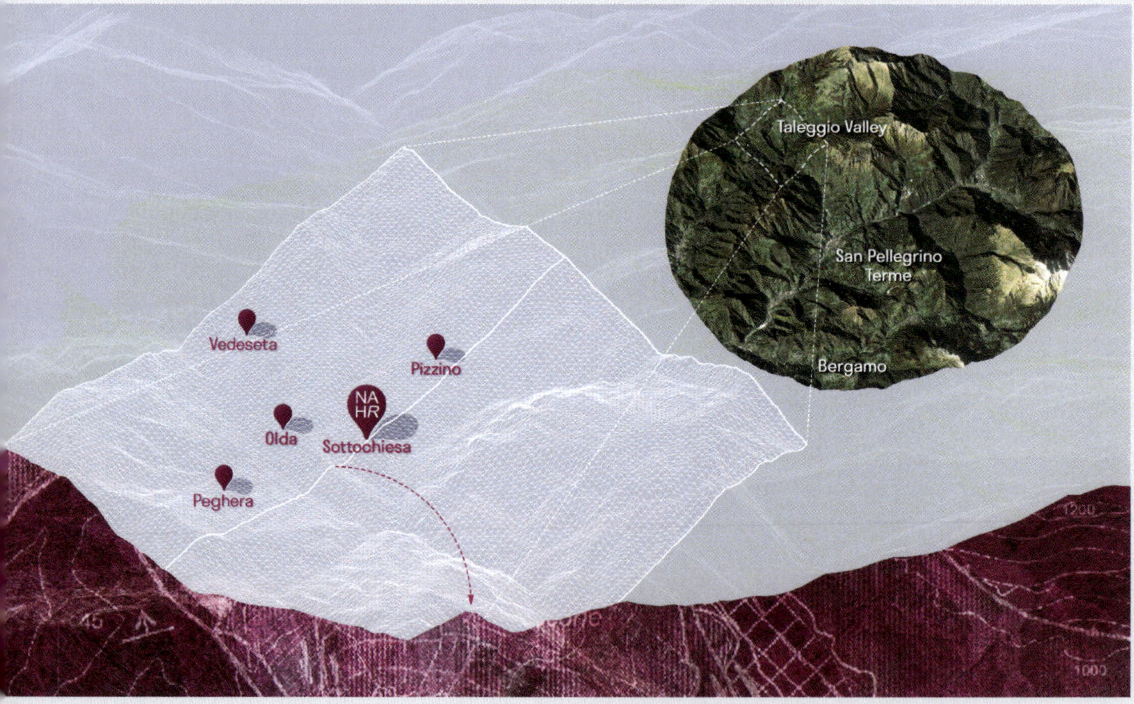

2 - Richard Molina. *Taleggio Valley Topography and main Villages location*, 2023. (Image courtesy of designer)

[3] The term biophilia here references the concept explored in E.O. Wilson's 1984 book *Biophilia*, where he defines it as "the urge to affiliate with other forms of life".

If this way of being sinks in, and exists in us, priorities will quickly shift. As human beings, we often concentrate on the idea of comfort in designing houses, cities, and landscapes, in order to conform to perceived needs. Perhaps shifting our comfort concept, together with limiting our abundant consumption and disposal of goods, is the first step we can jointly take to move towards the restoration of a resilient earth. How wonderfully fortunate it might be for all living organisms to be able to live in unpolluted areas. How delightful to envision that priorities of wealth will be strictly associated with neuro-psycho physiological health. How might we achieve these ambitions? We can acquire new consciousness by observing natural systems and processes, by living closer to nature, and by coexisting with other species so that we can embrace multifaceted approaches to sustainability through intentional acts. Creativity helps us understand and move faster toward such paradigms.

In order to study the natural world, which is what ecology is preoccupied with, we must recognize the interconnection and interdependence of all ecological elements. In doing so, we also recognize its intricacy, and the complexity of humans coexisting with others who are perceived by many as separate. Yet each one of us is so fundamentally entangled that our responses will determine the future of our own species' survival.

It is generally difficult to tackle ecological topics one at a time, as everything in nature is interconnected and it is very important to express and articulate any element in relation to the other. To keep the discourse comprehensible, some linearity and introduction of parts one by one is required for a book. However, this book is conceived as a place for transecting various topics, as the relationships between different aspects of the global and the local perform together in an interrelated and richly detailed score that makes up the natural environment in its multidirectional web. Eventually, everything becomes entangled, or simply ecosystemic.

We are using the word 'nature' to prioritize the role and function of every diverse element of the ecosystem, whilst also integrating interpretations of the term beyond the purely scientific, which relate to our cultural selves. In fact, if we were to speak more properly of pure science, we would prefer using 'ecology': the branch of biological science that studies the relations of all living organisms to one another and to their physical surroundings. Yet we have chosen to use the term 'nature' more often to recognize the intrinsic impossibility of removing the human, and our cultural interpretations, in any readings of the ecological systems. The subjectivity and actions we have been imposing on the planet are hard to eliminate at this point in time. By decentralizing the human position, respecting 'the other' without the prejudice of hierarchy, we embrace the more-than-human world as an essential approach to the post-Anthropocene. To foster harmonious coexistence, we must explore innovative avenues for comprehending and empathizing with our fellow beings: active engagements that nurture direct connection rather than exploitation help to cultivate a deeper sense of presence and respect for all life forms.

3 - Walking the Terrain. Marco Ferrari. *Building Space - Space Buildings*, 2022. (Image courtesy of architect)

Art practices can act as a connector and amplifier of environmental messages. Artists study contemporary scientific research and traditional ecological knowledge, conduct fieldwork, and use the knowledge and inspiration to create work that may inspire others to be more empathic and attuned. As NAH*R* Fellow, Marco Ferrari, has commented, "walking through a terrain already constitutes a first act of inhabitation: ground unrolls and, absorbing the action of the body, reveals unprecedented imaginaries and spatial possibilities" (2022). Exhibitions that have traditionally been within the walls of museums, are now more often engaging the outdoors, in installations that highlight nature's beauty and also carry environmental messages. These include observations of ecosystem interrelationships, and conservation and restoration provocations. The fragility of life on our planet, especially in relation to extractive human actions, is powerfully expressed by the visual, literary, and performing arts.

The events and installations within the community ethically involve populations that would otherwise rarely interact with art, and create a broad social interaction with multigenerational emotional involvement. Installations in open spaces resonate over time, remaining in the stories and narratives of the many that have experienced and participated in them. Creativity is a powerful tool of freedom and consciousness, and artists, through personal fieldwork, ideation, and community engagement, generate questions and responses to our greatest contemporary environmental challenges.

By being proactive and acknowledging the intersections between each issue, we gain a clearer vision of the world, its multidisciplinarity, the relationships between the human and non-human, the way it works on a variety of scales, and the extreme or subtle processes at play. When examining human habitats, we recognize that they offer more than just construction. How we support or put pressure on the environment is brought into focus in the expression of our material cultures. By learning more about the ways design and buildings work, either with or against the natural systems around us, we can develop an understanding of how to change our actions and approach. It is paramount to recognize the non-human right to habitat as we change the shapes and functions of the land, and focus on preserving and conserving ecological hot spots.

4 - *Coexistence* exhibition in Bergamo City, Italy, curated by NAHR. Emilio Fantin. *With*, 2022. (Image courtesy of M. Macconi)

5 - Open spaces as places for ideation and participation. Taleggio Valley, NAHR groups, 2021. (Image courtesy of NAHR)

A PLACE FOR IDEAS

The collation of materials in this book displays NAH*R*'s aims to produce collaborative and participatory dialogues which explore the multilayers and intersections of art and science, nature and culture. NAH*R* is a non-profit that was founded mainly by architects as a reaction to the environmental impact of humans. As architecture is at the intersection of, and in collaboration with, many other disciplines, ranging from the arts to engineering, we found ourselves needing to expand this discussion even further, to make it truly collaborative, horizontal, open and permeable, approachable and defensible by all.

NAH*R* presents multidisciplinary research methods and supports non-formal learning fieldwork. There is inherent value in the rural setting for this non-institutional think-tank, given its intellectual positioning as a globally-focused entity within a local, and historically rich, community. As such, it aspires to foster a place for ideation and co-creation that can be scaled, as appropriate, to other territories: a method at times, an example at other times, an experiment most of the time!

6 - Walk the Talk: Mountaineer and NAH*R* fellow David Chan talks with the grasses to discover their identity and present them to the world, while hiking in the Alps. David Chan. *Ground Truth: Grasses and Pastures of Val Taleggio*, 2019. (Images courtesy of artist)

There can be a deep mindfulness cultivated towards our environment, that is, to be in nature, to talk about nature and to 'walk the talk'. This means traversing the terrain of the Valley's grounds and immersing oneself in the ecosystems of the mountains. At NAH*R* we emphasize nature's insights as an instigator for thinking and fieldwork. Ultimately, we aim to unfold and display a sensitive type of culture that relates to nature as a source of informed inspiration and a measure of available resources. This can be achieved in novel ways by looking macroscopically to Earth's ecosystems–climate, geology, soil, water, air, sun–and then reading microscopically the Valley's natural and cultural characteristics, including flowers, ferns, bees, pebbles, barns, villages, commerce, and more. The creative minds' development of these topics and subjects is essential to NAH*R*'s participatory practices found in this book.

7 - Communal and cooperative fellows interact in meetings and workshops. Field explorations at NAHR guided by naturalists Cristina Movalli and Enrico Bassi, 2021. (Images courtesy of I. Mazzoleni)

LET'S PLAY!

Creativity is a potent tool for a deeper appreciation of the ecological realm. NAH*R*'s premise is to foster a multidisciplinary discourse that focuses on fighting climate change and biodiversity loss through creative actions. Publishing this book marks the ten year milestone of these activities. The complexity of these topics mirrors the ideas embodying our ambitious objective to transform coexistence and create an eco-centric world. We bring participants from all over the globe to the culturally and historically rich rural setting of Sottochiesa, Taleggio Valley, in the Italian Prealps, and since 2021, to the outpost in the Santa Ynez Valley, California. The Fellows include, but are not limited to, architects, landscape architects, visual, media and performing artists, writers, filmmakers, and scientists. Each year, they reflect upon a different ecological theme, and these have shaped the six chapters in the pages to follow. NAH*R* facilitates space and time to work on a specific elemental natural theme. There they can examine the resiliency of the systems, especially those connected to the various valleys in the region.

With the premise of our non-profit in mind, the authorship of this book had to be multi-vocal. Along with the authors, there have been multiple contributing editors, and all of the one hundred NAH*R* fellows have been invited to express their voices through texts and their work. Just as polyphony in music and literature can represent multiplicities, its dialogic nature holds here possibilities for change as many voices are heard simultaneously, in respectful harmonies. The participatory nature of NAH*R* highlighted in the voices, creates the polyphony necessary to address complexities of contemporary environmental challenges. There is hope for a sustainable resolution when we work collaboratively and collectively.

Moreover, to write about the Earth without imposing a human-focused instrumentalism on the discussion, we must understand humans as one of the actors in the natural sphere. This book pushes the discourse about our impact and connection with nature onwards, encouraging and inviting new paradigms through which to understand our habitats. We consider the ways we mine, occupy, inhabit, and coexist within the world through the lens of creativity, activism, and research. The book poses questions and solutions to drive us away from the extractive economy, which has been developed in line with the growth of all industries. Urban design, architecture and resource management, along with other fields, are key players in our search for a positive and non-destructive future relationship with the Earth.

WALK WITH US

Collaborative, participatory, and field-based work is critically urgent today. We are interested in practice as applied research, and in encouraging explorations which are neither purely academic, nor purely professional. There is a fast and productive feedback loop from discussing to exploring, to designing, to making or performing and testing. And there is the process of discovery and creation that is constantly in motion. The work here is thus grounded, hands-on, and in touch with communities of humans and non-humans alike.

NAH*R* aims to foster a space for participation, reciprocity, and inclusivity within larger environmental dialogues. It is an open space, not governed by institutional forces, available to all voices that not only want to be heard but are interested in sharing, as acts of both generosity and care, and responsive to listening or evolving, no matter the career path or stage of life. Towards developing an enriched network of thinkers, activists, and creatives, NAH*R* has reached beyond the Fellows to the kind and open heart of the local community first, and then, outwards to people and institutions of the larger region. Over time, by strengthening and multiplying the connections (also virtually), the community has become global, displaying the depth of cultural wealth that biodiverse rural locations can provide to the willing inquirer. While the Fellows take their thoughts away to their urban lives, there is left behind a distilled and increasing sense of value within the place.

On this journey towards clarity there will undoubtedly be long periods of 'lost in translation' as a necessary condition for wondering, noticing, asking and questioning what we do, and why. It is this fundamental uncertainty which gives rise to possibilities especially when pondered through multidisciplinary conversations and exchanges.

Anyone who attends or interacts with NAH*R* is asked to be open to partake and play, and consider the rural setting as a case study through which to explore in-depth the domains of nature and landscape. The locations' identity and memory are rich with biodiversity, and they hold strong material and social cultures within which the Fellows work. Approaches can include engagement with the tangible and intangible heritage of rural buildings and artifacts, as well as oral stories, local knowledge and know-how, traditions, crafts, flavors, and habits of the local population and communities that partake, legitimize, perpetuate, and generate an *in-situ* cultural logic of people and place. But most fundamental to NAHR remains the terrain one can traverse and transect. Walking the talk is central to our explorative, inquisitive attitude towards the ecologies of each site.

8 - Guido Pesenti Campagnoni at San Marco. Mauro Sacchi and Yu-He Lin. *we were here*, 2015. (Images courtesy of artists)

As we promote work inspired by nature's resilience, we have identified four key perspectives, or domains. These domains are not hierarchical and often intersect with each other, helping fresh transversal explorations to shape new perspectives. The first domain looks at the implementation of a *regenerative economy* adopting long term vision and systems thinking as fundamental tools in reaching resilience both in local and global production. Thus, there is potential for balancing extractive modes of operation with resilient and regenerative strategies, prioritizing the commons and the biodiversity of the ecosystem. One of many examples is the study of complex relationships of the grasses and pastures, such as the local flora, microbiota, cows, milk, cheese making and its commercialization, and how restoration of circularity amongst these agents helps eliminate waste and pollution while reinforcing the territory's diversity. The local precise exploration of this domain provides insights and models principles of regenerative economy applicable to different localities and even to scaling up in consideration of global systems.

The second domain focuses on developing *bio-inspired designs*. The local natural forms, functions, processes, systems, and cycles are providers of both the rich physical context for creative inquiry and also models of resiliency to guide a design practice (product, architecture and planning) beyond the dominant anthropocentric extractive *modus operandi*. Projects within this domain find inspiration in nature's solutions to coexistence. The aim is to lessen the environmental impact of human actions upon the natural ecosystems, and invite contributions to the commons: from the bark of a tree we can find inspiration for the design of building facades, while we can look at symbiotic relationships for enhancing gardens planning.

9 - NAHR's four domains. (Image courtesy of NAHR)

Another key domain involves thinking about how we design our futures by engaging with *technologies and new media*. For example, remote sensing to track the changing animal behavior in a particular area through time is used to understand the effects of climate change, while 3D scanning a quarry to understand its vast geological formation and environmental impact imposed upon all species. These are two examples of how new technologies can be of aid in advancing humans' comprehension of the complexity of our planet. The fast growing toolset provided by this domain is increasingly critical as it expands our possibility for dynamic observation of interconnected events, augmenting human engagement through multimedia modes of representation, and dissolving the boundaries between human-made and nature.

Finally, our own bodies offer the next domain, as primary tools to experience the world within and around us. Through *embodied investigations* in the realm of the performing arts, new forms of knowledge often emerge in which we are as part of nature as the trees, waters, rocks, or animals, breaking the boundaries and becoming at times one with the all. Using our bodies as apparatus for direct engagement with/in nature provides immersive sensorial and responsive trans-sectional multi-body explorations. Immersing oneself in a river, dancing bare feet with the rocks, or listening to soil by lying down on it have all led to powerful subjective discoveries which are otherwise not possible to uncover through a distant gaze. Authentic care for the environment is borne out of such discoveries. An extension of this domain also led us to experimenting with creative eco-writing and ecopoetry to reflect upon these experiences. Stories and poems have proven to be critical tools in the way we inquire about the elements of our surroundings and in the way we reconfigure them for restorative and healing stories. As Donna J. Haraway (2016, p. 12) has noted:

> It matters what matters we use to think other matters with; it matters what stories we tell to tell other stories with; it matters what knots knot knots, what thoughts think thoughts, what descriptions describe descriptions, what ties tie ties. It matters what stories make worlds, what worlds make stories.

POLYPHONIC PARTICIPATION

The book aims to give space and time to multiple points of views, therefore it is structured so that each chapter includes a short introduction framed by the chapter's editor, an essay on the global perspective, and another essay on the local context of the Italian Prealps, the Taleggio Valley and its region. To complete each chapter there are the *Voices* of the Nature, Art & Habitat Fellows who participated each year, enriched with a selection of their work from the archive: their reflections and provocations complement the two original essays. While each essay's linear reading is critical to capturing the depth of the content, reading the book transversely allows for numerous enhanced interconnections and expanded understanding of the climate crisis we are witnessing and contributing to, and how we can get involved and create new work that helps in shifting the current situation.

The six chapter topics are presented in the order in which they were explored at NAHR, beginning with *The Woods: Pathways to Reforestation*, a primary element of the Valley, representing a sustainable, renewable yet depleting resource globally. *Rock and Stone: Material Culture & Cultures of Making* represents the substrate of mountain life, giving shape to the natural landscape and conferring character to the vernacular built environment. *Water: Vital Flux, Energy in Transformation* unpacks water as a source of life and a means of access, energy, sustenance, and survival globally but also specifically in the Taleggio Valley. *Grasses and Pastures: Imagining a Regenerative Economy of Cheese* focuses on the socio-cultural and economic processes of manufacturing cheese in the Valley, and on the complex interrelationships with the environment and the regenerative economy that it generates. *Animals: Interdependence Between Species* highlights the animality of our own species–our inner and outer dialogues both among us and with others–as well as our necessity and struggle to coexist. Finally, *Soil: The Critical Zone* guides us through the complexity of the world below our feet which sustains us all but which we have also abused and depleted at a fast rate.

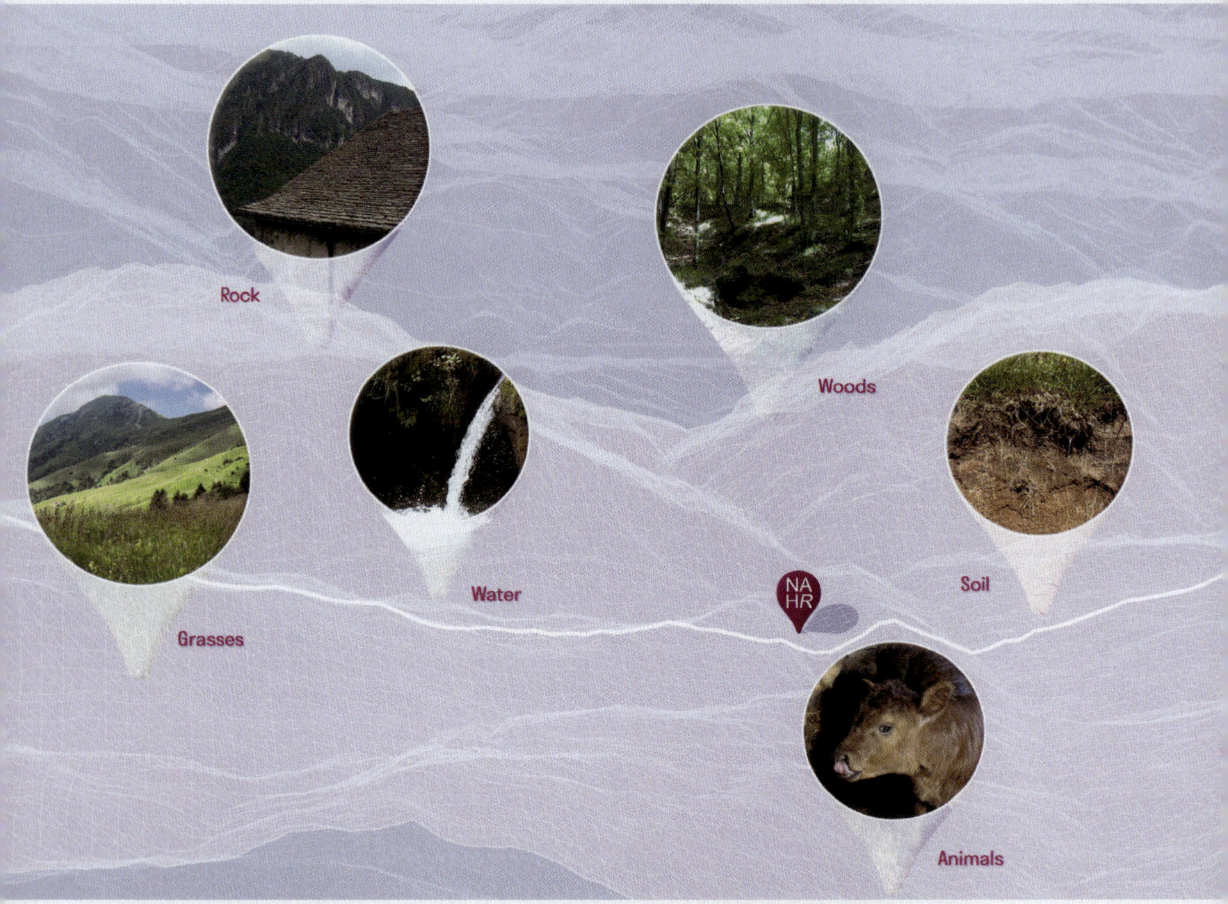

10 - Richard Molina. *Six Elements of Nature in Transect of Coexistence*, 2023. (Image courtesy of designer)

Without pretending to fully explain or dissect the complexities inherent in these interconnected elements, each chapter gives food for thought and a platform from which to learn about the urgency of our situation, while witnessing how others are proactively working to address it, and at times even proposing solutions! Each element brings together a larger community of voices that, in their totality, create a polyphony which resounds with a call for actions beyond words.

11 - Andria Zachariou. *Moving Memories*, NAHR EMPACT, 2023. (Image courtesy of artist)

NAHR'S ONGOING EXPLORATION

Why might you journey with us, dear Reader and fellow explorer? The collection of writings in the pages to follow seeks to provoke and generate new thoughts; to ignite new scholarship and inspire innovative approaches to living and making. The book offers an inquiry into environmental issues through multi-disciplinary, eco-centric thinking in structured, thematic areas of concentration. The multiple voices, stories of resiliency and precarity, represented together seek to unveil the intimate links amongst living organisms and elements of nature. By listening to nature and viewing it through the lens of others, we better understand the ways we can inhabit and restore our planet.

This book intends to invite a wide and diverse readership to engage in contemporary dialogues alongside scientific and creative research, and to rethink professional work and the governance of territories. The contents will satisfy the curious, presenting possibilities to expand knowledge systems, and most importantly, take action on whichever aspect connects and appeals most to one's own life. You will find snapshots of NAHR up till now. Explorations and observations of local issues suggest ideas and models for applications elsewhere in the world. For the scientific mind, this book might encourage more openness and sharing towards the general public as it highlights possibilities for interdisciplinary, cross-pollination of ideas. For the creatives, it encourages one to take on environmental subjects, to learn from and disseminate knowledge of nature and the systems that support all lives.

Transect of Coexistence: Inquiry into Nature, Art, and Habitat aspires to contribute to increased consciousness through a journey across viewpoints, topics, possibilities, experimental works, voices, opinions, and facts. Our contribution in the field is to pinpoint how our rich cultural heritage and fragile ecologies need to be seen as more interconnected, and that we need to learn from each other in order to reach the homeostasis necessary to allow life for all living organisms. Coexistence is not only critical, but essential to us all.

Acknowledgments

NAH*R* was founded in 2016 by architect and biomimicist, Ilaria Mazzoleni, in collaboration with those who first conceived and believed in the possibilities of such an initiative: researcher Anna Santi; anthropologist Alexandru Balasescu; advisor Alberto Mazzoleni; and visionary architect and designer Alessandro Mendini. As the NAH*R* community grew, the contribution of ideas from the core founding team expanded the activities of NAH*R* with determined efficiency, thank you: Gabi Scardi, Adrian Paci, Paola Tognon, Maria Cristina Rodeschini, Kyong Park, Chiara Gamba, Fulvia Mendini, and Dan Disney, joined soon after by ambassadors Melissa Blanco, Jose Herrasti, Noah Mercer, Lucinda Coleman, Sara Invernizzi, Aslı Suner, Richard Molina, Roxanne Korda, Doe McAndrew, Sofia della Pace, Anne LaForti, Elena Stevanato, Jenny Rodenhouse, Stefania Cabassi, and Katie de Bari. Others joined the evolving conversation and we wish to express genuine gratitude to those essential to the growth of NAH*R*: Deborah Weintraub; Maristella Casciato; Enrico Bassi; Paolo Belloni; and Susanna Ravelli. Each of these generous and talented members helped fostering new partnerships with institutions and agencies, including the Ecomuseum Val Taleggio, Associazione Amici Artista Franco Normanni, Australian Poetry, Fondazione Comunita' Bergamasca, Ordine degli Architetti PPC di Bergamo, EMPACT, European Union, Helms Design District, Northwestern University, Berlin University Alliance, University of the Arts London, Culver City Arts Foundation, Midlands4Cities, ArtCenter College of Design, CalArts, Abitare, as well as Associazione Santa Croce, Flora Alpina Bergamasca, Landscape Festival Arketipos, Orto Botanico di Bergamo L. Rota, Cheese Valleys le tre signorie, Associazione Casa del Bergamino, Consorzio per la tutela dello Strachitunt DOP, WWF Lombardy, Centro Studi Valle Imagna, Pro Loco Val Brembilla, and with professionals such as naturalist illustrator Stefano Torriani, author Cornelia Funke, environmental lawyer Paola Brambilla Pievani, governmental executives Gloria Cornolti and Lucia Mangili, communication specialist Angela Anthony, food writer Daniel Quintero, reporter Frances Anderton, and radio host Evan Kleiman. These partnerships have enabled relationships and contributed to supporting residencies initially in Taleggio (Italy), and since 2021 also in Santa Ynez (USA), hosted by architects Sarah Graham and Marc Angélil at their ranch, r3, to whom we are especially grateful. We thank each and every visitor and speaker that over time enriched our learning, whether during our in-person workshops and open studio days or in remote webinars. Many thanks to the extraordinary Fellows-in-Residence who took time to be responsive to the environments of the Taleggio and Santa Ynez Valleys, nourishing NAH*R* through artful actions and exciting discourse. Your work is valued; your time, energy, and creativity are gifts to our community - *grazie mille.*

NAH*R* began conceiving the idea of this book years ago, in collaboration with select individuals whose innate goodness made this publication possible. The guest writers who contributed global and local essays invested considerable time, demonstrating generosity of spirit and passion through their written contributions. Each of the NAH*R* Fellows, whose visual, poetic, and written thoughts shape the essays entitled Voices and punctuate the global and local narratives, have enriched the book through pithy insight, sudden bursts of beauty, and rigorous art-making as meaning-making. External contributors such as photographers and illustrators who have allowed use of their images have also given vital visual material that illustrate our authors' works. You can read the biographies of each valued contributor in the final pages. We also thank our publishing team, ListLab and Alessandro Martinelli, for their agility and expertise; to the colorful April Greiman for her cover design; Malena Martinez for her poetic and expertly insightful foreword; and book whisperers, in particular Anne Elvey. The driving force behind the book has been the NAH*R* team who donated their time and expertise. Thank you in particular to the editorial team: for your incisive wisdom, Maristella Casciato; dancerly writing, Lucinda Coleman; native knowledge, Sara Invernizzi; outstanding commitment, Roxanne Korda; exceptional direction, Ilaria Mazzoleni; simple brilliance, Richard Molina; exquisite calmness, Gabi Scardi; thoughtful enthusiasm, Aslı Suner; and your patient guidance, Deborah Weintraub.

We must also thank the generous people of the Taleggio Valley and vicinity who have been supporters, collaborators, and participants in all of the NAH*R* stories. In particular, we acknowledge and express our deepest gratitude for the support of the municipal administrations of Taleggio, Val Brembilla, Vedeseta, and the Municipality and the Province of Bergamo. We have many other institutional partners to thank, such as: Regione Lombardia; Comunità Montana Valle Brembana; GAL Valle Brembana 2020; UNCEM; CAI Bergamo; Università degli Studi di Bergamo; Parco delle Orobie Bergamasche; Val Grande National Park; AIAPP Lombardy; and BIM Bergamo. The incredibly dedicated Chiara, Veronica, Marta, Maurizio, Franca, Battista, Monica, Giuseppe, Rosangela, Osvalda - and others - of the Ecomuseum Val Taleggio have given invaluable support to NAH*R* activities over time. Similarly, we are grateful to Alessandra, Paolo, Barbara, Agostino, and other friends for their participation over the years. During the past ten years we have had the support of the farmers, breeders, artisans, entrepreneurs, professionals, historians, restaurateurs, the Parish, and the whole population, from the young to the elders, of the Taleggio Valley. You are too many to name individually, but we deeply thank each and all of you.

Thank you, Fabio Scatena, Yarita Escobar Sori, and the entire Mazzoleni family, starting from the *Soggiorno*'s late founders that dedicated their existence to the well-being of the place NAH*R* now calls its home. The new generations with Alessandro, Alicia, Gianmaria, and Maya as continuing participants in several NAH*R* activities. Sottochiesa, its territory and history, has become the serious playground hosting many of the Fellows' works, though in general the entire Valley and the Prealps have been instrumental to the success of NAH*R*. We want to pay homage to our ancestors, to the valleys and the mountains, to the flora and the fauna, and all sentient beings for whom this land is home!

And thank you, dear Reader, for your engagement with the material of this book: we hope you are inspired and energized by the makings and musings of the words, within.

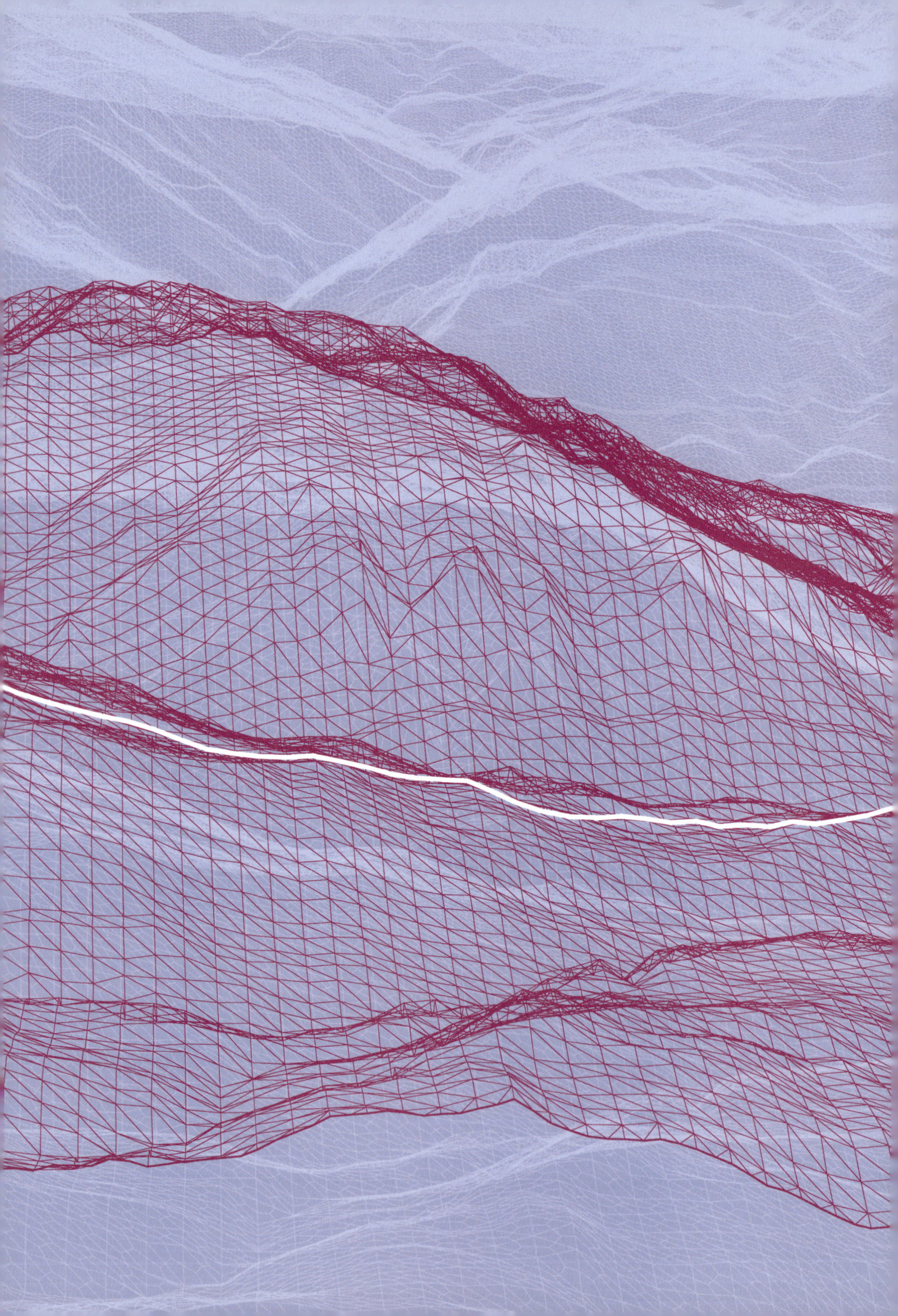

CHAPTER 1

Woods
Pathways to Reforestation

Introduction

Maristella Casciato

Trees, trees, trees. . ."We often think of trees as objects. . .victims of their own immobility. . .They grab on the planet, as [the world-renown botanist] Anthony Huxley put it. . . able to regenerate, to begin again from the beginning. . . We forget that they breathe every day, leaf by leaf." (Leonardi & Stagi 2019). With their by now classic handbook *L'architettura degli alberi* (1982) Cesare Leonardi and Franca Stagi opened the path to the consciousness of trees as living species, well before the escalation of the climate emergency and the design call for sustainability. Nowadays, the demand is to produce strategic thinking and technical knowledge to counteract the fact that the relationship between the architect and the tree has not always been one of sincere appreciation. Indeed, to be malicious, one might discern in it a certain prolonged distrust, as if the former envied the latter their spontaneous facility to create elegant forms which, additionally, became more and more apparent over the years if allowed to be shaped by environmental conditions.

The precedents of a dialogue not deprived of ambiguities are illustrious and centuries old, from Plato through Leon Battista Alberti, who in *De re aedificatoria* took pains to consider why those spaces that in the ancient city were considered suitable for "children to grow more robust in contact with the open air". When considered as an object of imitation, he recognized their origin in a tree-lined garden as modern sensibility would also seem to suggest to Palladio.

Yet, it was the naturalistic spirit of the Enlightenment that flourished in the eighteenth century that reminded us that the relationship between trees and architecture was substantial. Although, not even the conciliatory attitude of Abbot Marc-Antoine Laugier, who rooted architecture itself and her orders in natural forms, led to definitive results. In the following century, the great theorist, Camillo Sitte, a cantor of urban beauty, returned to question the actual appropriateness of placing trees alongside the most majestic buildings. Finally, in the twentieth century, it was fully understood that the city had changed and that the urgency of the relationship between construction and nature was not so much about the distant past, but rather addressed the future to come and the very ability of architecture to be accepted by its users.

Nevertheless, if human-colonized territories of woods can be said to be marked by a historical imprint, this is also reflected in forests which have been deeply and pervasively influenced by human activities. Compared to the evidence of the anthropogenic landscape of urban areas or agricultural lands, the presence of human modification activity in a forest is more difficult to recognize. The forms taken by forests over time depend on the various ways in which they have accompanied the changing socio-economic conditions of places: in relation to their association with the grazing of wild animals, the production of timber for construction, firewood and charcoal, shipbuilding, or fruit production, to mention a few of these appropriations.

The cultural motivations for this age-old distrust, at micro-and macro-scale alike, are precisely embedded in the terms we use to define these environments.

"Wood" comes from Old English *wudu*, which means "trees collectively, forest, grove." Sometimes, "wood" was generically used for "wild" as opposed to "domesticated."

Conversely, the etymology of "forest" comes from the Latin *foris* and did not mean forest or woodland at all. In fact, the ancient Latin term for this environment was *silva*, from which the English language derived the term savage, that is, wild.

On the other hand, the Latin term *foris,* which gave rise to "forest," means "outside" or "out of doors"; the term also being used to refer to the door and the related threshold. Hence, it is understood that from *foris* came the English term "foreigner" and the adjective "foreign", and in reference to places, "outside the boundaries of a country", or to persons, "born in another country".

The forest continues, therefore, to be a foreign, remote, unfamiliar, unrelated, unheard-of, often unfriendly place.

The etymology thus testifies through the oldest roots of the terms *forest* and *wood* to the consideration of the foreignness of those environments to human presence. But local traditions, anthropological research, and socio-economic history applied to the environment lead us to different considerations, as the two essays in this chapter evidence.

Luca Mangili cannot initially escape from the myth of the forest as "nature sanctuary", *lieux* of romantic encounter to borrow Pierre Nora's vocabulary. Yet, it is the reality of the present that forces him to regain attention to the need for recreating that biodiversity that could guarantee the duration of the forest, excluding, if necessary, the consumer. In conclusion, his plea returns to the mantra of positive values, compatible with safeguard and ensuring the protection of biodiversity.

From what could not be a more distant approach, Andy Lipkis takes the tree as the object of his action. For decades he has been an advocate of the tree "acupuncture" within the urban fabric. Trees equal medicinal substances that produce healthy effects and allow the cycle of life to remain vital. Trees collaborate to create urban forests.

It is worth noting how the forest, at a global or local scale, is the goal where most of the environmentalists keep their faith for a more sustainable planet.

As so, in recent years, the inclusion of densely wooded areas in cities, under the seductive name of "urban forests" or "urban woods", with the aim of creating islands of climate mitigation and biodiversity enhancement, have again aimed to combine two seemingly distant worlds: the natural world of the wild forest and the human-made world of the individual home, namely the city. A relationship between the forest and human being has always been marked by an intrinsic duplicity: the recognition of the forest as a foreign environment, but at the same time its indispensability which the climate emergency poses with dramatic urgency.

In the "Voices" section that closes this chapter, the authors introduce new language when they write about the way in which everyone becomes a "creative partner with nature." Hence the appeal by NAH*R* Fellow, Isabelle Duvivier, "Don't shove trees where they fit in, put trees where they need to be and change the infrastructure to accommodate the critical nature so it can thrive".

Trees Need People and People Need Trees

Andy Lipkis

Take a deep breath. Hold it, and count the seconds before you need to exhale and take another breath. Humans can live only six minutes without oxygen. We can live only six days without water. We can live only six weeks without food.

—*Richard St. Barbe Baker*, attributed by author

Let us consider tree planting as an act of *acupuncture*: a deliberate, studied and precise action that can be used to heal an imbalance or wound. Trees can treat and heal a number of problems, such as floods, pollution, loss of habitat, air filtration, etc. However, as with acupuncture, the planter must make sure to place the correct tree, in the correct location. Even the planting of one tree needs careful consideration in order to avoid a cascade of effects on the local ecosystem, such as disruption of the habitat, restructuring of vital air and water flows, and creation of fire hazards.

Human actions such as urbanization, deforestation, pollution, and resource exploitation, have deeply wounded our towns, cities, woods, forests, and wider environment. These self-inflicted wounds have compromised the ability of the environment to provide for and protect humans and other living organisms. As will be discussed, by starting with an individual sapling, or grove of trees, we can work towards restoring a whole woodland, the surrounding nature and ecosystem services, and even work towards creating lived-in urban forests.

Trees supply all living organisms with the key elements for life on Earth. The air, water, food, and herbs required by all life forms, including trees, plants, medicinals and soils, animals and people, are produced, cleaned, sustained and/or recycled by the ecosystems of trees, plants and the soils in which they grow.

Trees, woods and forests have a beautiful, fundamental, and co-dependent partnership with humans. But as human populations have exploded across the planet, our partnership has become complicated and broken. Most people living in densely populated, developed, and urbanized areas no longer grow and harvest their own food, or personally collect and manage their water and waste. We no longer directly experience our impact when consuming or harming nature and natural systems, nor do we experience working in partnership with natural resources. Thus we have lost sight of the need for reciprocity; returning and recycling the energies and resources provided by the trees and forests so they can continue to provide for, protect and support our very lives. This is a fundamental part of our role in this partnership.

The dislocation from our relationship to the ecosystems, trees, and forests that support us, has been deepened by market-based economies, through the mining of nearly everything, including every human *experience,* such as emotions, love, attraction, beauty, self-worth, and the *perception* of power. Value is determined by what the *market* will pay for the 'resource'. But there are externalities which are not part of the market's accounting. There is no value placed on ensuring resources are healthy, sustained, renewed, cared for, or even acknowledged. The full extent of their essential ecosystem, climate, health, and life support services are not identified or understood. The result is our natural resources are starved, down-cycled and consumed, without reciprocity. Resources that were naturally recycled and upcycled by the ecosystem are instead starved of nutrients (chemical, physical, emotional and spiritual) and are no longer able to fulfill their essential and unique roles in providing the ecosystem services necessary to ensure the preservation, restoration, and sustained availability of those resources necessary for life on Earth. Thus, in order to meet our own needs, humans now have a key role in replacing and sustaining tree life, from forests to towns.

We often underestimate the importance of urban forestry and tree management in an increasingly urbanized world. When we see the trees we are inclined to notice their above ground individuality and beauty, proudly standing in their own spaces. However we rarely consider their need to be part of a collaborative network, both above ground (through birds, insects, and the airborne pheromone signals and seeds) and underground through the soil, water, and roots. Without the aid and collaboration of humans there is little space for these connected tree communities to form in the urban and constructed environment.

TREE ECOLOGY - A BRIEF OVERVIEW

Trees convert solar energy into sugars that feed them, so they can grow and make leaves, fruit, seeds, and fiber (wood) which feed, heal with their medicinal herbs, and make habitat or homes which support most life on Earth. In turn, many of these elements fall to the ground, decompose and make more soil and nutrients that feed the trees and well as other plants, fungus, bugs, micro-organisms, animals, and people, enriching the ecosystem and perpetuating the cycle of life.

We know trees breathe in the CO_2 that people and animals exhale, along with the carbon gasses that are produced in decomposition of organic matter and combustion of fossil fuels and wood. And we benefit as trees convert the CO_2 into oxygen so we can breathe it in. Again. And again. Trees 'fix' or lock up the CO_2 in their wood, leaves, and in their roots and the soil in which they grow.

1 - Decomposing organic matter in the woods. Annalisa Zegna. *WETLANDS research project*, 2022. (Image courtesy of artist)

Trees and forests also 'manage' all the water on Earth, critically important in urban areas. They capture and slow down the rain in their leafy canopies, and in the moss and grooves along their branches and trunks, then carefully release it into the very porous soil held by their roots. The spongy soil and root zone acts like a tank and water treatment system to hold and process massive amounts of rainwater. The soil sponge is comprised of soil, decomposing and composted leaves, mulch, and a community of billions of microscopic organisms, fungi, bugs and critters who live there, filter, and clean the rainwater before it recharges groundwater aquifers or runs off and gathers to form rivulets, creeks and rivers, ponds, lakes and oceans. Water evaporates from the leaves in the trees' canopy and recharges clouds and the atmosphere with water vapor that carries moisture and rain to other areas in what once was an unending cycle that continually watered lands across the earth.

The tree and its root zone are a major component of Nature's flood protection system, and can support urbanized areas through creating a preventative barrier within or surrounding a city or housing district. For example, the root zone under a single California oak tree with a 100-foot-wide canopy can capture, store, and treat 120,000 gallons of water in a 12-inch flash flood. Imagine what happens when that one oak tree is removed. The water rushes off, taking precious soil, rocks and pollutants with it, flooding, contaminating and eroding areas and potentially killing people or damaging homes, roads and towns downstream before it eventually flows into the ocean. Now imagine what happens when dozens, hundreds, or thousands of acres of trees are removed. The soil and its capacity to re-grow trees can be lost for hundreds or thousands of years, alongside the increased risk of flooding and contamination of the soil and waters. If properly selected, placed, planted and cared for, trees can and do play a significant role in rainwater capture and groundwater storage, as well as pollution and flood mitigation, in urban environments.

2 - The oak tree canopy is 100 feet wide. Its soil is a habitat for hundreds of organisms. Its canopy catches rainwater and slows it down. Dirty water flows to the root catchment with a capacity of 120,000 gallons in a flash flood. Water is cleaned and then infiltrated into the aquifer. After this the tree evapo-transpires and recharges the atmospheric moisture, 2023. (Image courtesy: concept Andy Lipkis, illustrations: Katura Reynolds)

The health of the planet's forests is greatly interconnected with the wider environment. Some climatologists, including Goldman Prize winner, Michal Kravcik, say that a significant amount of ocean sea level rise is caused by runoff from deforestation due to land clearing and paving for urbanization, not just the melting of the polar ice caps (Kravčík 2007). This wide scale clear-cutting of forests and destruction of watersheds has far reaching impacts beyond our urban centers.[1] It is accelerating desertification, warmer temperatures, and water cycle disruption and can include unforeseen negative feedback effects of healthy forests elsewhere. Heroic human actions to reforest and catch water from previously cleared forest land, has shown dramatic 'healing' results. Rajendra Singh, a doctor in India, was awarded the 2015 Stockholm Water Prize for his organized efforts to restore the traditional rainwater harvesting functions on cleared lands, through tree planting, and creating small catchments. These have resulted in restoring seven separate rivers, as well as the failing agricultural economies of local villages (Water Stories 2022).

Above ground, the urban tree canopy plays a major role in mitigating air pollution and extreme heat, alongside providing a critical habitat for biodiversity. Insufficient urban tree canopy has been implicated as a culprit in the creation of urban heat islands, inner-city areas without shade, which experience a compounding of heat with much higher peak temperatures than surrounding suburban and rural areas. These urban heat islands have been thought responsible for the significant increases in human mortality during heat waves (Heaveside, Macintyre & Vardoulakis 2017). In Australia, and the United States, public health officials began collaborating with urban foresters to dramatically increase urban tree canopy cover to save human lives after experiencing very high numbers of human fatalities during heat waves in the early 2000s. The trend is extending to European cities as well (Lungman et al. 2023).

OUR ROLE AND RESPONSIBILITY TO TREES

Observing humanity's profoundly consumptive and destructive behavior, one might conclude that all non-human life on Earth would be better off without people. But what might be lost to trees, forests, and all life forms without the presence of

[1] The areas where rainwater falls and flows to a creek or river are known as a *Watershed* in some countries, and in others, they are known as *Catchments*.

people? Let us imagine the Earth itself requires the presence of both humans and trees as tools to maintain a balanced system. To understand why this is the case we would look to the full set of functionalities and capabilities of humans and trees respectively and together. We have created comfortable urban environments that, for the longest time, have separated the built environment and Nature. Only recently have we started to realize that no living organisms can exist in isolation. Certainly, within urban environments our actions have gone so far that we now see the need for some of these global relationships to work in reverse: trees may now need people for their survival.

Everything on Earth has unique capacities that enable the performance of functions and actions that are absolutely critical to the health of the entire planet. We must recognize our role as caretakers in time to prevent a full collapse of the ecosystem services that enable humans to live, thrive and survive. So how can we deploy and engage with this in urban settings?

Humans have highly evolved capacities to sense, respond to and solve problems, or heal pain and suffering. We are equipped with brains, hearts, and guts that each contain *neural receptors* which enable us to sense, amongst other things, pain, suffering, and brokenness (Armstrong 2017). As individuals we are all attuned to different issues and callings. There is however a fairly wide array of issues that we are all sensitized to, having to do with living things, such as other people, animals, and elements of our communities and the environment, that we know or feel to be important to our survival, health, and quality of life. When something threatens or harms any of these things, our neural receptors notice and message us with adrenalin, causing us to wake up, pay attention, respond, and engage. An example is what happens when you see or hear someone who is injured or calling for help.

This response has been labeled 'fight or flight', but that profoundly undervalues and misidentifies human nature (and in fact all living organisms' natures) and our innate capacities. We all have the ability and need to repair, heal, and grow things (Wade 2009). Paying attention to, and responding to these 'calls for help' develops a vital strength, connecting us, and showing us the joy in helping others. That is true, grounded, and natural power.

Our biology has us innately tuned in to animals, trees, forests, rivers, and other aspects of the environment, but without direct connections, relationships or knowledge of them, we may not distinguish the call, or be able to respond. We will often feel disturbed and agitated by the triggered adrenalin leading to unexplained anxieties. Nonetheless, we have been crafted and evolved by the ecosystem to solve problems and humans are absolutely called upon and needed to help. This can be through a process of restoration allowing their ecosystem services to thrive and function in all habitats, from our own neighborhoods, to cities to damaged forests.

Urban activities leading to air and water pollution, global warming and drought have damaged forests not only through direct deforestation but also through lowering the trees' natural disease and pest defenses. In many evergreen species, when bark beetles attack and bore into healthy trees, their sap, or 'pitch' as foresters call it, flows to the wound and pushes the invading beetles out of the trees. This has been observed from the conifer forests in North America, to Europe, Siberia and beyond (Katz 2017). But when weakened by air pollution, the trees' protective sap-pressure is catastrophically reduced, allowing the beetles to burrow into and eat their way up and down the cambium layer just under the bark, where the living soft tissue and vascular system conveys water and nutrients through the tree from the roots to the leaves and needles. The beetles lay their eggs in the burrows, and when the eggs hatch, the young larvae eat the cambium tissue all the way around the trunk, completely cutting off the flow of nutrients and water, killing the tree that had grown there for hundreds, or even thousands, of years. During the summer of 2022, massive fires burned hundreds of thousands beetle-killed trees across the Boreal Forests of Alaska.

In many cases, such forests can no longer regenerate on their own, and in this period of increasing climate change and threatened ecosystem collapse, the need for replanting and re-establishing the native trees and forest ecosystem is both urgent and more challenging. It is vital to prioritize protecting and re-establishing the fragile and sometimes fraying web of life so that all endemic (naturally, locally occurring) species of plants, animals, birds, insects, and fungus can support the integrity of the global ecosystem and natural migration of species.

There are large efforts proposed and in some cases, already underway, to use drones and airplanes to drop millions of self-sowing seedlings in biodegradable tubes. When the priority focus is simply establishing any species of trees that will grow and sequester carbon, even if they survive, they may be counterproductive if they are not endemic or naturally occurring in that bio-region. One important way to ensure the highest chances of take-up from these seedlings is by informing and guiding the selection and planting of keystone species for dispersal through traditional indigenous knowledge and guidance from Indigenous peoples. Actual humans, not robots or drones, are needed to plant and look after, tend, mulch, weed, water, protect and monitor the trees so they become established.

SEEDING LIFE

One important factor determining the ultimate success or failure and health of the trees as they grow up is determined by their seeds. Seeds for growing seedlings should be sourced as close as possible to the planting site. Indigenous seeds are genetically programmed to survive and thrive in their unique environments and microclimates. This includes factors such as elevation, slope, aspect (the direction the slope is facing), hours of sunlight, minimum and maximum temperatures. Ideally these will nearly match the conditions of the source and planting area. Significant departures from the matching microclimate may result in the newly planted trees not receiving enough light, water, or viable growing temperatures, resulting in unhealthy, weak, stunted, spindly adults, or simply dead trees.

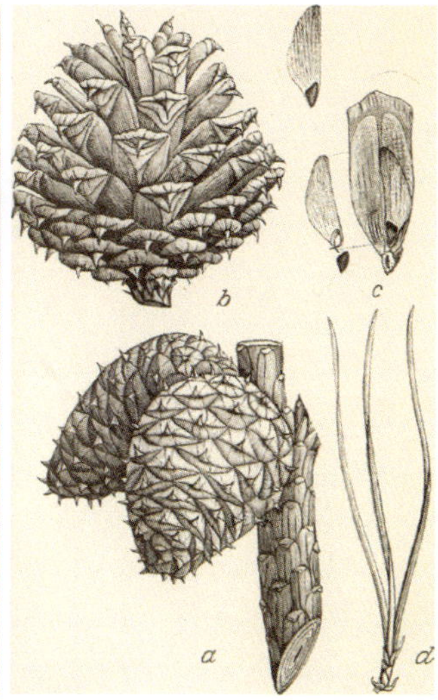

3 - Famartin. *American sycamore* seeds along *Tranquility Lane in the Franklin Farm section of Oak Hill, Fairfax County, Virginia*, 2022. (Image courtesy of Creative Commons) Pinecone seeds and needles, George B. Sudworth. *The forest nursery: collection of tree seeds and propagation of seedlings*, 1900. (Image courtesy of The Commons) Roger Culos. *Coast Redwood Cones and Seeds*, 2015. (Image courtesy of Creative Commons)

If the design and species selection aren't grounded in such knowledge, and ongoing human monitoring, the essential restoration of ecosystem functions is questionable and it is hard to foresee long term success in the development of supportive relationships between humans and trees.

The following case studies, showing examples of the interactions between trees and humans in urban and domestic environments, will highlight situations in which developing a respect for the value of trees has had clear and positive impacts on human experience.

The Work of Treepeople in Los Angeles, California

Five decades of pioneering experience in leading urban and rural tree planting has illustrated the need to engage large numbers of people in a process from planning to planting, to long term care, monitoring and adaptive management.

4 - The Mighty Quill. *Pine beetle damage viewed from Mount Fraser,* 2006. (Image courtesy of Creative Commons)

In 1973, I launched an effort to reforest areas where air pollution from greater Los Angeles had weakened and killed millions of trees in the mountains and National Forests one hundred miles east of the city. After the *Los Angeles Times* shared the story of my efforts to bring thousands of seedlings to youthful summer campers, thousands of people donated to support the effort. By providing environmental educational briefings on the state of the forest and the need for their help, young summer campers across the forest planted trees, and watered them throughout the hot and dry summer. Everyone we worked with called us 'The TreePeople', and ultimately we accepted, formalized, and adopted the name.

We monitored the tree survival rate over time and found that the trees which received water, mulch, weeding, and several years of care, survived. Without intensive human care, many of the trees didn't make it. The same is true in cities and *urban forests* across the United States. The US Department of Agriculture study reports survival rates of trees planted on American public streets to be between 19 to 28 years: rarely enough time to establish a broad canopy to shade and protect people from the *urban heat island effect* (Allsup 2020). The heat island causes the chemical propagation of air pollution and compromises public health. In many large cities, including Los Angeles, heat waves increase mortality for people with chronic diseases who live in disadvantaged communities with under 6% tree canopy cover.

TreePeople learned that deep engagement with people—learning, planning, and investing their ideas, dreams, and personal energy in planting, watering, caring for and protecting their trees for at least the first five years—resulted in much higher survival rates.

In one major project, a TreePeople-trained volunteer *Citizen Forester*[2] dreamed of a large-scale project to organize and support community members across South Los Angeles in planting a living monument to Dr. Martin Luther King along the seven-mile-long boulevard named in his honor. On Dr. King's birthday in 1990, 3,000 people joined hands along both sides of the entire street and planted 400 trees in three hours. Most of the trees are alive today because of the human relationship and partnership that was formed with each of the trees and the entire boulevard.

5 - Andy Lipkis at the same spot on Martin Luther King Avenue Los Angeles in 1990 (A) and 2008 (B). All trees planted and cared for by "citizen foresters". (Images courtesy of author)

Each tree was planted by a team of three or four people assisted by a pre-trained volunteer. Once each hole was dug and the tree placed into it, the planters were invited to pluck some of the hair from their head and place it in the hole, so the tree would ultimately 'eat' their hair and absorb their DNA. Each tree was named by the planters, and a neighbor living within 100 feet was recruited to care for the tree, and given a special watering bucket.[3] The name of the tree, along with the planters, and the adopted caregiver were all inscribed-on vinyl bumper sticker material and affixed to the steel tree stake (support pole). Upon completion of each tree, the planters joined hands and chanted, "trees need people and people need trees," and welcomed each tree by name. But that wasn't the end of it. Every month, for ten years, Jim Hardie, TreePeople's Director of Forestry, organized and supported tree care, weeding, and watering with at least 100 community members. Now, 30 years later, most of the trees are alive, and tall, and form a green line through South Los Angeles that can be seen in aerial photos (Accelerate Resilience L.A. 2023).

[2] A retired school teacher named Eudora Russell.

[3] The same approach has worked on a massive scale in Nepal, where millions of people participated in replanting and long-term care of their new trees in their deforested mountains. It's a rare, and powerful lesson. The reforestation and ecosystem restoration work requires human caring and human energy to succeed.

Native Trees, the Oak

Since founding TreePeople, I have always been delighted to hear how the seedlings we supply have fared. Before delivering a keynote speech at the *UCLA Economic Forecast* a woman thanked me for providing perhaps two dozen of coast live oak seedlings twenty years prior, which she had planted around her hillside home near Malibu. With a tone of urgency, she continued, "I want you to know those trees saved my home in the Woolsey Fire. The fire came down the canyon above my house, and the trees just stopped it—they didn't ignite. Later, after the fire had burned the hills around my home it came back up the hill from below, and again, the trees I'd planted below my house and on the sides, just blocked and stopped it from reaching my house. Nearly all the other homes in my neighborhood were destroyed in the fire, but the oak trees saved my house" (pers. comm.).

The coast live oak trees she planted are indigenous to Southern California and thus are equipped to survive and support and protect life in that climate. Once established, their trunks and branches are too thick, dense and moisture laden to ignite in a fast moving wildfire.

This story is emblematic of the vital importance of the often hidden relationship between trees and people and our potential for sustaining life and health, and perhaps turning the tide on climate collapse. This is a story of hope, love, creativity, partnership, respect, understanding, learning, reciprocity, and adaptive inquiry, for humans and Nature.

6 - *Coast live oak tree*, 2013. (Image courtesy of I. Mazzoleni)

Indigenous knowledge of the multiple gifts, benefits, and ecosystem services provided by trees is extremely relevant and important to respect when informing how to take action. Grounded and guided in part by this perspective, architects, scientists, engineers, artists, and activists are studying and learning the climate protective functions and natural engineering of native trees, forests, and watersheds (or catchments) and have *bio-mimicked* their functions (such as the oak and its water capture) to inform the retrofit and adaptation of city and suburban infrastructure.

Another Example of Urban Forestry, Australia

In Australia, where hundreds of people have died in extreme heat waves, scientists discovered that increasing tree canopy cover in their cities to 40% was sufficient to lower peak temperatures enough to save lives. Both Melbourne and Sydney launched successful efforts and campaigns to increase tree canopy cover, to 40% and 50% respectively (Coutts et al. 2013, 2016; Jamei & Tapper 2019).

The research and implementation in Australia inspired similar studies in Los Angeles, which revealed that massive discrepancies in tree canopy cover matched significantly higher mortality rates in its extreme heat events. Los Angeles has launched a tree canopy effort to redress the issue (Salabert 2022).

Drought, flooding, and fire confound and worsen the challenge. Traditional urban design principles aim to keep water and people apart, to protect homes and neighborhoods from flooding, and standing water which breeds mosquitos, and can rot buildings. But this well-intended approach has worsened water shortages during droughts, as cities lose the precious rain that does fall.

During a nation-wide and decade-long drought starting in 2000, Australia launched an intensive effort which biomimicked the natural watershed functions. They created a massive campaign to financially support the installation of rainwater harvesting tanks to catch and save water for use on their trees, gardens, and other critical needs. In their major cities, as much as 47% of the homes installed large rain tanks, which enabled them to survive and save many of their trees (TreePeople, 2016).

Back to the USA, the Parkways

We have also seen a positive exploration of tree watershed functions in relation to flood control in the USA. In the cities of Seattle, Washington, and Portland, Oregon, the traditional flood protection approach was discovered to be the cause of the near extinction of seven species of salmon throughout the region. Engineers and ecologists teamed up to redesign the distribution of trees and foliage in their street-adjacent parkways, to bio-mimic the forest watershed ecosystem starting from the creeks, so rainfall could be captured, slowed, filtered and bio-remediated (cleaned) of pollutants, then safely returned to the local creeks where the salmon were able to once again successfully spawn.

In 2010, Los Angeles began implementing this approach, in a very hot suburb appropriately named Sun Valley. The intention was to prevent severe flooding, clean pollution from the water, and make it available to support new vegetated bioswales and trees and recharge the city's underground aquifer (TreePeople, 2012).

Multiple studies conducted by TreePeople, the City of Los Angeles, and the University of California Los Angeles (UCLA) have shown that the city, which today imports most of its water from other regions, could supply fully half its water locally by capturing the rain that falls on the city. That can be accomplished by widely deploying the aforementioned approaches of biomimicking forests and watersheds across the city. Such an effort could employ many thousands of people, especially chronically under-employed and unemployed youth, along with the active voluntary engagement of people across the city. Currently, pumping ⅓ of LA's water over the mountains and into the city is the single largest use of electricity in the State of California, which is the fifth largest economy in the world. A substantial portion of the effort could be financed by replacing fossil fuels that are used to pump water into Los Angeles from distant forests, with human energy. Imagine instead, the implications of paying people for human energy to plant, grow and care for the urban forest watershed to capture, clean and store the water and create a safe, sustainable and resilient future.

We can see in these examples the importance of remembering and respecting traditional indigenous knowledge, making use of native species and reconstructing the forest systems that had originally evolved and adapted to that land. Through deeper understanding of the processes of nature, and the use of trees to aid in the healthy flow of resources, we can make a better life for both humans and trees, particularly in our urban environments.

CONCLUSION: WHERE ARE WE NEEDED AND HOW WE CAN HELP

Ultimately whether one lives in a large city, town, village or in the countryside, we all have opportunities to take small but potent actions to build resilience that will grow over time. It is important to ground ourselves in *three* fundamental cornerstones of approach that will ensure the success of your efforts by building foundational knowledge.

First: *Work* respectfully with the living elements of your local ecosystem to discover what's needed. Ask local experts, or learn about the woods, forest, watershed and habitat where you live, before it was altered by farming or urbanization. When choosing seeds check with local foresters and silviculturists to understand the optimal growing conditions, limitations and sensitivities of any intended replacement tree species. An example of inquiry, deep respect, and partnership with Nature and the ecosystem is detailed by the Native American botanist, ecologist, horticulture professor and author Robin Wall Kimmerer in her book *Braiding Sweetgrass*.

Second: *Learn* about the profound foundational role of native plants in re-creating a 'beach head' for restoration of the vitality of our local ecosystems, from attracting and providing habitat for native birds and wildlife, to growing a source of traditional medicinal herbs and remedies. Diana Beresford-Kroeger, a renowned botanist and biochemist, details a simple approach for rebuilding climate resilience in her book, *To Speak for the Trees: My Life's Journey from Ancient Celtic Wisdom to a Healing Vision of the Forest.*

Third: *Share* the vision. A successful tree planting effort is strongest when built on a shared vision, including tools, experience, knowledge, connections, funds, and long term care such as watering, weeding, and feeding. Without at least five years of sustained care, the chance of trees thriving and surviving profoundly diminish. As the climate changes extreme weather events are increasingly overtaking the designed capacity of infrastructure and public services. Emergency services can become overwhelmed and are often underprepared, but when neighborhoods and communities connect, dream up, design and implement a tree project together, they unwittingly build the most resilient infrastructure. If a storm, fire, or flood strikes, you want to know that a neighbor will feel connected to you and safe enough to wake, warn, or rescue you, and your family. Without these connections, we can experience unnecessary and unfathomable loss. This community process is laid out in the freely accessible text *The Simple Act of Planting A Tree: A Citizen Foresters' Guide to Healing Your Neighborhood, Your City, and Your World* (Lipkis & Lipkis 1990) which guides the reader in enrolling, engaging, and building a team of neighbors and friends as they work together to implement a successful tree planting project. The book has helped to grow the 'Citizen and Community Forestry Movement' worldwide.

Plant a tree. Save a life. Enjoy. Celebrate. Repeat. When we care for trees in this way then communities, of plants and humans alike, can support each other to live safer and healthier lives.

The Woods and Us:
Between Resilience and Interdependence

Luca Mangili

Talking about the woods means considering different points of view and sensitivities. In the imagination of many it is the natural place *par excellence*, but for others it is above all an economic resource, and these two perspectives often lead to conflicting attitudes. Ancestral reverence for the forest survives in the unconscious imagination: a wild space to enter with respect and amazement, but the idea of the woods as a nature sanctuary is relatively recent, and we must find more concrete and shared motivations from the community to effectively communicate the irreplaceable value of biodiversity. At the same time, the economic consideration of the woods cannot be reduced to the simple rotation of the cuts and the quantity of timber collected, or to the banal use of the adventure parks. There are obvious repercussions on our well-being and on sustainable production processes when we do not value the woods' important ecosystem services. Services generated by natural systems in favor of humanity are generally divided into four categories: in support of life, for supply, for the regulation of the environment, and for the defense of cultural values. Whether protected or exploited, how we perceive the woods impacts our changing relationship with nature, which has evolved over the course of history and always needs to be given new meaning.

A LOOK AT THE PAST

The very idea of the woods contains in itself a drawback. The woods are nothing but the impoverished fragments of the immense primary forest that completely covered the plain, and compacted upwards through the valleys and mountain slopes to their upper climatic limits, which have descended or ascended several times.

About 6000 years ago, between the Po and the central Alpine arc, the dynamic succession of forest types manifested itself fully, finding full expression also in the Bergamo area (Lombardy). The oaks, especially the English oak (*Quercus robur*) dominated the plain unchallenged, while the oak (*Q. petraea*), the Turkey oak (*Q. cerris*) and downy oak (*Q. pubescens*) also prevailed at hilly and low-mountainous altitudes; the former on more fertile and fresher soils, the latter on poorer and drier ones. With the increase in altitude, the beech (*Fagus sylvatica*) gradually took over, favored by greater rainfalls all year around, even during the summer months. In the mountain range, the beech forest was fully established, sometimes with a discreet presence of silver fir (*Abies alba*). It then gradually transformed into the spruce forest at subalpine altitudes, where

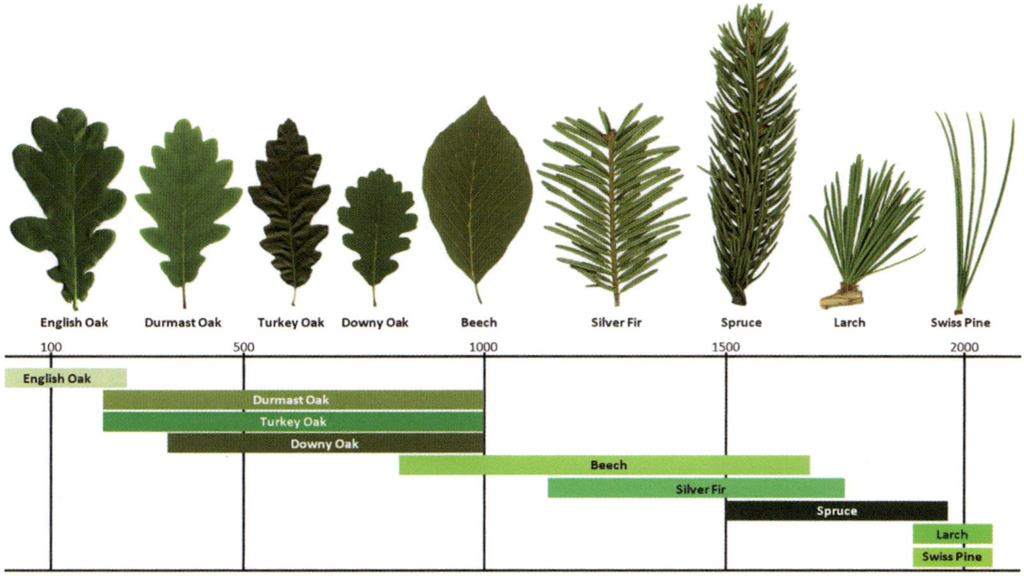

1 - Diagram of the vegetational altitude bands, 2023. (Image courtesy of author)

the progressive rigidity of the climate suited the spruce (*Picea abies*). Further up were the larch (*Larix decidua*) and Swiss pine (*Pinus cembra*) which constituted the extreme offshoot of the forest.

This primeval succession has been upset by humankind. For millennia, deforestation has been a usual practice which has accompanied the expansion of civilization. The current woodlands of the Alpine arc and the Po valley have been repeatedly cut, thus constantly reduced to make room for crops, meadows, pastures, and settlements over the centuries. Many woods have suffered fires, have been replanted and grazed, and reconstituted. People have favored some species to the detriment of others, causing profound transformations.

From the second half of the 19th century, wealthy citizens discovered the mountain as a place for recreation and a romantic encounter with nature. By the following century, the birth of mass tourism extended this romanticized notion of relationships of mountain populations with the woods. People behaved as if they had reached a mature awareness of the importance of the protection of the environment and the quality of the landscape, having imagined the woods in prematurely conservationist terms. Yet there remained sensitivities foreign to them. In reality, it was substantiated very differently, because the economy of pure subsistence meant that those who practiced removals and reconfigurations could not ignore its intensive exploitations.

The photographs taken almost everywhere in the Alps from the end of the 1800s to the early 1960s show very small wooded areas, confined to the steepest slopes, in the gullies and on the rocky slopes that could not be used for other purposes. Wherever possible, meadows and pastures were obtained and became the main engines of the mountain economy. The residual woodlands provided timber for work and, above all, for burning. The production of charcoal was also important as it was indispensable for pre-industrial metallurgy.

2 - Reggetto area (Val Taleggio): comparison between the orthophotos of 1954 and 2018 (SITer@ Province of Bergamo); the expansion of the bosco and of the built-up area to the detriment of the meadows is evident. (Image courtesy of Regione Lombardia)

In many woods there are still numerous and well recognizable *poiat* (local term) which are small flat open spaces. On these spaces, a circular pile of wood was expertly arranged and covered with soil for seven to ten days, then due to partial combustion, this *poiat* became coal. Here and there you can also glimpse the remains of the charcoal burners' shelters, forced to live long periods in the woods to constantly monitor the various *poiat's* in business.

In the lower valleys, the chestnut groves were particularly cared for, as a kind of 'domestic' woodland. While in the hills, the chestnut grove was managed as coppice, to satisfy the huge demand for poles for the vine. Introduced in Roman times, the chestnut was later systematically spread at the expense of oak groves, with

3 - *Aral* with reconstruction of a *Poiat*, 2014. (Image courtesy of author)

4 - Chestnut dryer in the Lombard Prealps, 2019. (Image courtesy of M. Macconi)

which it shares an ecology. The chestnut was defined as a providential tree for the nutritional value of its fruits, and numerous varieties were selected, with gradual maturation and intended for different uses, then propagated by grafting on wild specimens. The chestnut was also highly appreciated for the excellent quality of its wood, as durable and suitable for almost all uses in the peasant world.

AND IN THE PRESENT

The spontaneous recovery of the woods that has occurred in recent decades does not derive from a different sensitivity, but from the partial abandonment of the traditional hay-making and grazing activities, which allows it to regain part of the original surface. The reforestation process, even though at first maybe hard to spot, becomes quickly visible, within a decade. However, the simple reconquest of the lost surface is not sufficient to restore the original conditions. The action of human intervention has been invasive and prolonged. Thus, it should come as no surprise that the current woods are to varying degrees different and impoverished compared to the original forests, being the only ones which were in total harmony with the climate and the soil and also free from human interference. Not even the woods that today appear more intact have escaped the current consequences. Often, rather than the tree component, it is the shrub layer and the herbaceous undergrowth that reveals its past and natural quality.

Today the plain retains only small portions of the ancient forest mantle, which still feel impressive when the oak tree numbers are appreciable: in the best examples, some surprising presences in the undergrowth testify to a largely lost floristic richness. Woods of no value are more frequent, heavily polluted by alien, often invasive, species

5 - Degraded woods: the black locust has completely replaced the English oak, present with only one specimen, 2010. Mixed woods with a very natural appearance; in reality the spruce was introduced in the primitive beech forest, 2014. Rare example of lowland oak forest with trees of different ages and rich undergrowth, 2011. (Images courtesy of author)

of Asian (*Ailanthus altissima, Ligustrum lucidum, Broussonetia payrifera*) or American (*Robinia pseudacacia, Acer negundo*) origin. The underbrushes are heavily destructured, they present a discontinuous tree layer with a dominating chaotic undergrowth. Evolution towards a return of an oak forest is blocked by anthropic disturbances, and the plains are a dangerous source of floristic pollution for the entire surrounding areas.

On the hillsides and low mountains, the oak groves have been almost totally replaced by chestnut woods. Despite a modest revival of interest, the groves are in a state of conspicuous abandonment and strong degradation due in part to epidemic diseases. They are now dominated by the exotic locust tree, which vegetates late and provides little coverage, allowing the luxuriant development of brambles (*Rubus sp.*) and elderberry (*Sambucus nigra*) to the detriment of the more valuable species.

On sunny slopes, on poor and dry soil, centuries of coppice management have strongly penalized the downy oak compared to the black hornbeam (*Ostrya carpinifolia*) and the manna ash (*Fraxinus ornus*) which are both able to grow back from the stump more easily; the resulting formations have a characteristic physiognomy, sometimes vaguely Mediterranean.The beech woods remains the prevailing wood type in the mid-mountain, even if its original extension was significantly reduced. In the past, the forest was intensively exploited for the production of wood and coal, and it is often polluted, or partially replaced by spruce even at lower altitudes. Yet, everywhere there is good evidence of the capacity for recovery.

6 - Beech woods in Val Taleggio, 2015. (Image courtesy of author)

7 - Spruce woods of the same age and poor in biodiversity, easily attacked by bark beetles, 2010. (Image courtesy of author)

For the spruce woods the discussion is more complex. The area that is climatically ideal, has given way to pastures. Yet, the growth of these woods has increased considerably due to the mismanagement of the past. Unfortunately, the short-sighted reforestation implemented for decades has always favored the spruce planted at very low altitudes and in unsuitable ecological contexts. The result has been extremely monotonous and fragile woods of the same age, very poor in biodiversity, unable to naturalize and renew themselves spontaneously, today aged and often perishable. The timber is in a dramatic crisis due to the very heavy attacks of the small bark beetle (*Ips typographus*) which quickly kills those trees already in precarious health conditions. The damage caused appears considerable, but significantly, highlights the incongruity of monospecific man made forestation, and calls for a re-evaluation of the spontaneous reforestation processes and a necessary attention to the ecology of the woods.

More generally, in recent years the stress conditions of many woodlands have appeared evident. Early vegetative growth is mortified by sudden returns of cold, leaves dry out and fall early due to drought, uprooting follows strong storms and parasitic attacks limit the biomass production. Accompanying these come the inevitable repercussions on ecosystems. Considering only the economic aspects, the lower productivity of the woods should translate into lengthening the cutting shifts, otherwise the resource will be depleted.

NEW SENSITIVITIES

Since the 1950s, the prevailing attitude of the population towards the woods has undergone a substantial change and there has been a shift from a purely utilitarian consideration to a more nuanced one. While not excluding economic exploitation, priority is given to the protection of the natural environment alongside a view to its touristic recreational use, though both outcomes are not always compatible.

In many hilly and mountainous areas, the downsizing of agroforestry-pastoral activities has severed the functional ties between settlements, crops, meadows, pastures, and woods, which were once legible in the landscape and often translated into toponyms with a clear meaning. In these contexts, the use of the woods mainly satisfied a local demand, while only a minor share of the timber was directed towards the capital city and the centers of the plain. Today, however, the management of the woods in a predominantly local context and aimed only at the extraction of resources no longer appears desirable.

8 - Centuries-old chestnut woods in the Lepontine Alps, where fantastic narratives of witches and spirits and the signs of abandonment have settled over the years, 2023. (Image courtesy of S. Invernizzi)

Citizens, who are unaware of agroforestry-pastoral activities, often have an idealized vision of the woods and nature, which they too easily consider uncontaminated. While they are better informed about global problems, such as climate change and the loss of biodiversity on the planet, there is a lack of understanding of the impact of global issues on local ecologies.

In recent years, it has been observed with concern the drastic decrease in rainfall and a consistent rise in average temperatures, with repercussions on water supplies, agricultural production, energy consumption, and quality of life.

The short and long-term effects of climate change contribute to increasing awareness of the ecosystem services provided free of charge by the woods. These benefits include the production of oxygen and the retention of pollutants, with beneficial effects on air quality; the storage of a large quantity of carbon dioxide which is the main cause of the greenhouse effect which causes the temperature to rise; a balancing action on the climate, through the mitigation of sudden changes in heat and the regulation of the water cycle; pedogenesis and provision of nutrients for vital biochemical cycles; protection from hydrogeological instability, which is even more precious now that precipitation often occurs with sudden phenomena of great intensity.

In the past, the practice of reforestation conceived the woods as a plantation of trees aimed at the production of income, with little consideration for the natural aspects. Today, the attention for biodiversity leads us to consider it as an ecosystem based on the complex and delicate balance between a multitude of producer and consumer organisms, which intertwine relationships of mutuality, competition, symbiosis or parasitism. Whether they are plants, animals, fungi, or bacteria, they are all necessary for maintaining the state of health of the woods, which is a necessary condition for ecosystem services to be dispensed on a regular basis.

It is the complexity of the natural woodland ecosystems that guarantees their duration over time, which did not happen for the monospecific planted woods. However, the exact appreciation of biodiversity is not within everyone's reach.

Except for the owners of the meadows and the mountain pastures, the majority of people evaluate the spontaneous recovery of the woods positively and believe that over time it is destined to evolve towards increasingly natural forms, close to those of the primeval forest, to the full advantage of biodiversity. This simplistic vision underestimates the value of polyphyletic hay meadows and pastures, which are environments of anthropic origin, rich in species and endowed with a high naturalness, comparable, and often superior, to that of the woods. In many areas, the impoverishment of biodiversity is directly linked to the disappearance of meadows, because grassland species do not survive in the woods and are destined to become locally extinct.

9 - Polyphyletic lawn for mowing, 2013. (Image courtesy of author)

In a phase of strong expansion of the woods, it should be assessed on a case by case basis to what extent to support and direct it, not forgetting that in the majority of instances it can occur completely spontaneously. Moreover, at least a part of the available resources should be destined to the maintenance of the most relevant grassland habitats for biodiversity. In reality, the evolution of new woods towards an ideal original forest is only theoretically possible; recently reconstituted woodlands always reveal the previous anthropic conditioning, ancient or recent.

10 - Jenny Rodenhouse. *The Enchanted Forest*, 2016. (Image courtesy of artist)

The condition of areas that have always been wooded is different, even if subject to traditional uses. Natural evolution can lead to better results, even if the primary forest will not reconstitute itself. Biodiversity is often greatest in transition areas at the edge of the woods, or in the clearings created by the fall of trees.

SUSTAINABLE HARVESTING AND COMPENSATION FOR ECOSYSTEM SERVICES (PES)

The assessment of natural quality should have the utmost consideration to distinguish income generating woods (in which periodic cuts can continue) from the old woods which should be protected in order to conserve natural habitats: a clear ecosystem guarantee. Thus two trends in apparent contrast are outlined, or experienced as such by individuals private or public bodies—who exercise the management of the woods. Sometimes the roles present as antagonistic, particularly when it is absolutely necessary to overcome the opposition between private interests (the search for high profitability) and the collective (protection of the territory and ecosystem services). The Regional Landscape Plan (PPR) and the Regional Law on soil protection constitute the regulatory framework and guidelines that should govern the interventions, guaranteeing the protection of the common good.

However, even large-scale considerations seem to distance the different positions. The demand for renewable energy which is exacerbated by the difficulty of finding imported fossil fuels, strongly relaunches the exploitation of the woods. In contrast, evidence of the climate crisis encourages their conservation and, where possible, increased growth. Furthermore, the global demand for wood is largely satisfied by the precious tropical primary forests, whose products also arrive in Italy. To safeguard them, would it not be advisable to increase withdrawals from secondary 'forests', which are our woods?

Sustainable forms of management, limited only to income generating woods, would still guarantee their existence. However, it should be noted that Italian production can only marginally influence the demand-supply-price dynamics. The various interests should find a solution in the Forestry Management Plans (PIF) issued at the provincial level and in compliance with regional and national regulations. The PIFs also consider the protective function, the importance of the landscape and the role of the woods in ecological networks. It appears there is the will to give new impetus to the forestry sector to encourage the rational harvesting of timber and the establishment of specific supply chains for its use.

The documents, created with care, quantify the surfaces and identify the woodland typologies present in the area. For each, there is objective evaluation of both the current conditions and the future prospects. The interventions suggested from time to time are all acceptable: withdrawals not exceeding the productivity, limitation or prohibition of clear cut, incentive of parcel or selective cutting, conversion to high forest species, de-setting, elimination of sharp contours in cuttings and in reforestation. These aid spontaneous regeneration in particular by favoring the entry of broad-leaved trees into artificial conifer consortia, and ecologically compatible native species in reforestation, as the precocious senescence of spruce plantings carried out in the past in unsuitable locations is now evident.

Despite good intentions, numerous critical issues have emerged. Coppicing, even by clear cutting, remains a recurring practice, because it facilitates logging, which can be problematic with parcel or selective cuts, and for this reason is often uneconomical. The satisfaction of family needs or the local demand for firewood still prevails over the quality production destined for sawmills. The foreseen phytosanitary and thinning cuts were not carried out, leaving the vast planted spruce woods to degenerate without the desired replacement with broad-leaved trees. The construction of agro-forestry-pastoral roads sometimes appeared dictated by the pure desire to take advantage of the funds allocated at various levels to satisfy particular interests, rather than by an actual need.

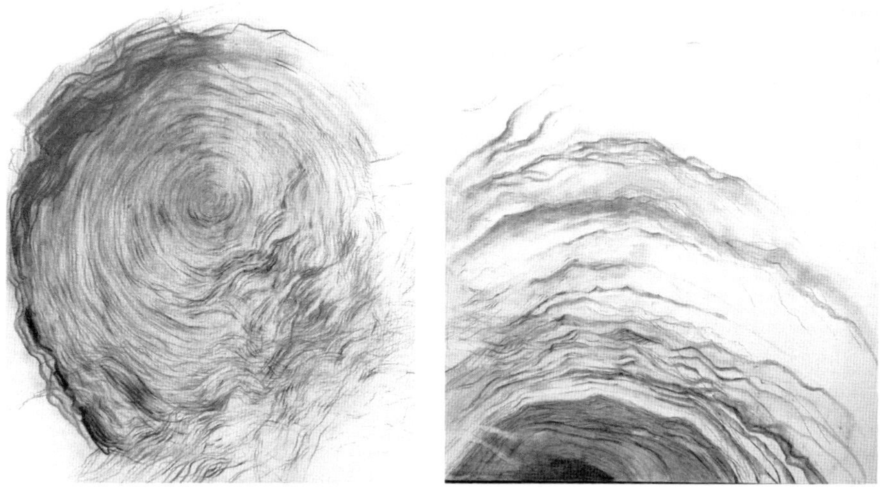

11 - Sarah Valeri. *Wood, Water, Air*, 2023. (Image courtesy of artist)

Furthermore, the PIFs are not the right tool to propose the remuneration of ecosystem services. There is a crucial element of connection between environmental resources and economic systems, which has been compromised by the prejudice of prioritizing economic outcomes over protection of the environment. Monetization is not simple and requires the development of reliable parameters and methodologies, but above all a forward-looking, agreed *governance* by public institutions.

Evaluations of a purely economic nature cannot have priority at the expense of the quality of the environment. The health of the woods affects the quality of the air we breathe, the availability of water, and the regulation of the climate. Management that privileges the achievement of profit by compromising people's quality of life is an absurdity. For example, in 2005, a Veneto municipality issued permits for collecting mushrooms in the woods. The fee generated an income that was five times greater than the sale of timber, and demonstrates how alternative approaches can preserve ecosystems and generate sound income.

In some Italian regions, and also abroad, there are examples of remuneration for the water management service rendered by woodlands. Owners are given a share of the tariffs for the distribution of drinking water. For example, in the city of New York, an agreement is active between the company for the supply of water services and the woodland's owners of the catchment area, who have committed to a management that guarantees continuity and quality of the water outflow. The surcharge recognized to them did not constitute a burden for the user, because at the same time a substantial saving on purification costs was achieved. In the Piedmont region, Italy, a variable quota from 3% to 8% of the tariff is destined to finance the activities of defense and protection of the mountain territory. A similar project was proposed in 2006 for the Val Nossana, which is the main water supplier to the city of Bergamo, but this was not implemented.

In the future, the carbon credit market, which is still embryonic in Italy, may also have interesting potential for development. The woods capture large quantities of CO_2, which can be certified and sold to companies that intend to work for the benefit of the global climate through tradable credits, each of which is equivalent to a ton of greenhouse gasses absorbed as a result of environmental protection projects. Obviously, ecosystem services must also be remunerated for the woods intended for integral protection, and in this way too these woods become a source of tangible income.

Faced with potentially promising developments, there are others that degrade the woods by reducing it to a sort of amusement park; a pure backdrop for recreational activities, such as the creation of adventure parks or downhill trails. Even recent regional provisions that allow free access to woods roads and paths with motorized vehicles for recreational or sporting purposes, as well as holding competitions, blatantly contradict the purposes of the PIFs and the Regional Landscape Plan.

12 - Saprotroph fungi decomposing the woods, 2020. (Image courtesy of author)

This defiance demonstrates total disregard for the value of woodlands and biodiversity. The regional legislation explicitly states that activities must be oriented towards the protection of the natural conditions. Unfortunately, the permitted behaviors cause direct damage to the trees and the undergrowth, facilitate the entry of alien species, compromise spontaneous regeneration, damage the soil by compacting it and causing erosion phenomena, become a source of pollution, and cause disturbance and the removal of the fauna.

The obstinate persistence of some stereotyped beliefs must also be considered negatively, such as considering conifers an obligatory and valuable component of any mountain environment, or the belief that the woods, to be healthy, must always be cleaned up like a park, ignoring completely the importance of the undergrowth and the cycle of organic matter.

CONCLUSION

The declarations of principles, the regulations, and the actions foreseen by the regional laws and by the PIFs appear adequate to guarantee the sustainable use and protection of the woods. However, they are effective *only if you consistently pursue* change, and in the pursuit of political consensus such practices are often disregarded. Moreover, greater consideration of the purely naturalistic aspects is appropriate for the identification of the most valuable woodland and grassland habitats so that their high biodiversity can be preserved and the ecological corridors between the mountains and the plains, once intimately connected, might be strengthened.

The wooded areas are considerable in the hills and mountains and very small in the plains. The extension of the woods is desirable, though not through new plantings on agricultural lands which is always insignificant in terms of flora and generates a depressing artificial appearance. Rather, the intention should be expansion, if possible, by connecting the rare surviving historic woods. In any case, it is advisable to create permanent grassy strips, which separate them from the crops where chemical fertilizers, insecticides, and herbicides are used.

If the sustainable exploitation of the woods continues in the hills and mountains, in the plains (where natural habitats are somewhat reduced and perpetually at risk) priority must be given to the protection of ancient woods. Protection can be facilitated by encouraging ecosystem services, which are currently not defined and yet critical for provisions in the future.

Safeguarding the woods and fostering sustainable use are compatible objectives, but require adequate knowledge to ensure that the biodiversity, history, and culture of the places are also protected. Such positive values contribute to the well-being of individuals and the community within the environment. Like all living beings, we too benefit from it, but we have a duty to contribute by spreading knowledge and practicing responsible behavior; the future of the woods depends on all of us, and our future...on them.

Voices of the Fellows
Woods

Lucinda Coleman, Editor

Woods are populated by individual trees that together create a diverse community. Similarly, this essay is made up of various voices from practitioners that have studied and devoted thought to the ideas and realities of the Woods. The invited contributors here were participants-in-residence of the *Woods*. Through inquiry-based practices, each Fellow responded to the power of forms and structures within the woods, and their voices now resonate with the residual traces of such encounters. As American architect, Isabelle Duvivier noted, "Instead of using the term *forest management*, we all need to think of ourselves as nature's partners, which means we need to become much more nature literate".

The following artists, architects, thinkers, and makers who spent time in this Prealpine Valley articulate their discoveries in written prose, through imagery, and in fragments of poetic text. Responses to the materiality of timber resulted in *made* objects; affective embodiment led to the making of ethereal, sometimes ephemeral, artistic artifacts. The following thoughts invite consideration of our own interconnectivity with the woods; a gentle provocation to connect with the trees, shrubs, and lichen within the woods of our own urban environments.

1 - Caroline Clerc. *Woods in Val Taleggio, Italy*, 2016. (Image courtesy of artist)

SEEING THE TREES FROM THE WOODS

Creative and/or artistic practices have the potential to point to, imagine, and produce forms of collaboration with non-humans based, for example, on reciprocity and mutual care. Perhaps a first step towards this is the recognition that much of our everyday life is dependent on or intertwined with the life of trees. In everyday objects, such as a table, dwells the life force of the forest.

A few years ago, thanks to Aymara Yatiri, I learned about the concept of *ajayu*. To explain it to me he took me to the Sunday street market of El Alto in La Paz [Bolivia]. There, looking at used vehicles for sale, he explained to me that *ajayu* is a vital force that resides in all things that exist, even in an old trunk. And that this is so because all the materials and forces that made something exist come from the same place, our earth. The concept became familiar to me, I understood that *ajayu* is present in various ways in Latin American culture, even in urban places. From then on, I became interested in perceiving and remembering the origin of things that surround me, with the earth.
Viviana González, Artist

3 - Thanos Zakopoulos. *The Valley Woods in winter*, 2023. (Image courtesy of artist)

A great part of the things in our daily lives, such as objects and even the air we breathe, depend on the life of trees. Consequently, our life depends on their life and our well-being on theirs. We are at the same time responsible for their welfare.

In each table inhabits not only the tree but also the soil that nourished it, the beings that inhabited it, and in general the network of relationships that made it exist. In each table, there is also a human force that transforms a tree into wood and wood into an artifact.

Awareness of the beings, forces, and relationships around us invites us to mutual care and reciprocity. Art and creative practices, in general, have the potential to invite, to share with collectivity this exercise of awareness.
Viviana González, Artist

2 - Collection of branches with a diverse range of sizes, colors, spurs, bark, scars, and ridges. Jay Yowell, 2016. (Image courtesy of architect)

4 - Viviana González. *Inside*, 2016. (Video still courtesy of artist)

LISTENING TO THE WOODS

You will literally hear the air and insects, sometimes water, wind, birds, and rustles unknown.
Caroline Clerc, Artist

Through entering silently into the woods, I am able to glimpse how the vital force of a community of trees may be and grasp some of this vitality that seems to be sleeping in our daily objects. When away from the woods I looked for pieces of wood in street rubbish and did further listening exercises (using microphones and interacting with those objects). The sound qualities of wood, such as resonance, are determined among others by the amount of moisture still present, reminding me in a vivid way of the moisture felt in the woods.

5 - Idil Kemaloglu (manufactured with master carpenter Stefano Offredi). *Kronos - Mountain Moruk*, 2015. (Image courtesy of artist)

6 - Mauriah Kraker. *Track / Shul*, 2016. (Image courtesy of NAHR archive)

Listening makes us understand how two moments and two states are woven into each other, to understand the complexity involved in the presence of a tree at a table. In a pile of wood fragments found in the streets of Bogotá, how many trees and forests would be present?

The sounds of the woods are murmurs that speak to us of that in-between space in which we and other beings might meet.
Viviana González, Artist

7 - Surya Tüchler. *Sketch for performance "RedMossStones"*, 2021. (Image courtesy of artist)

UNDERSTANDING THE WOODS

I guess exploring the limits of human language can be an interesting way to explore the limits of all kinds of languages. We could wonder, for instance, how a forest expresses itself, what are its gestures or its character, what can we perceive and point out about it?
Viviana González, Artist

I begin with making artwork and constructing images that explore the relationship between humans and the environment. Language follows.
Caroline Clerc, Artist

Roberto Esposito has long sought to interrogate what happens to notions of 'community' when bio-politicized performers operate to the interpellations of hypercapitalism. The theorist lands on the claim that the term community historically entails offering "to give up one's most precious substance, namely, one's individual identity, in a process of gradual opening from self to the other" (Esposito 2013, p. 84). In our era, Esposito rightly finds capitalistic social structures to have turned us each into "immunitarians," who stage atomistic, self-investing performances as if positioned outside of the community (Esposito 2011, p. 6), motivated solely by precepts of "'proper,' 'private,' and 'immune'" action (Esposito 2013, p. 88).

Within these global ideological contexts, can we really speak logically or critically of trees performing within their own 'communities'" or in doing so do we unwittingly transfer implicitly human analogues onto non-human living entities? Does this anthropocentric figuring once more centralize humans and perpetuate an insidiously instrumentalizing Master/Slave relationship? It seems that the ethical responsibility of any early 21st century creative producer remains explicitly tendentious[1]: we must strive to find ways to create capacious new lexicons, giving shape and form to newly ethical representations of possible inter-relational connections which entail, foremost, that we discard our in-built 'desiring' (exploited/evoked systemically by capitalism) in order to participate in post-capital communities in which nature's value remains valorized outside the logic of use or exchange.

8 - Dwelling surrounded by trees. (Image courtesy of I. Mazzoleni)

In short, in commanding the willing sublimation of 'self' in relation to others, we must do nothing less than set out to achieve the impossible. And the so-called 'tree communities'? Whatever these are, leave them to their own devices. Leave them as our revolutionary analogues, to learn from, at a distance, as if an order of visible, untouchable deities.
Dan Disney, Poet

[1] viz. Walter Benjamin's notion of the term, as found in his seminal 'The Author as Producer'

A MINDFUL APPROACH TO LIVING WITH TREES

Partnerships begin with mutual understanding and trust. And understanding requires knowledge. Creative practices can be a method to share this knowledge, which can lead to an understanding of the benefits tree communities have to human communities. I often think of how James Cutler, an architect in Seattle, honors the wood utilized in his buildings. He adopts a Native American approach to using resources. When something is taken from nature, one should honor its life. Cutler creates beautiful buildings, highlighting wood (rather than covering it up behind walls). When I begin a new project, I am interested in protecting the trees during construction and how the building may impact future tree growth. These trees give so much to our buildings (shade, beauty, free mulch, etc.), that we should be a partner in allowing them to thrive.
Jay Yowell, Architect

Ancient people, prehistoric wild grazing animals, and farmers have been coppicing trees and shrubs since prehistoric times. Coppicing is when young stems of trees are repeatedly cut to the ground making new growth at the stump. Coppicing has the effect of providing a rich diversity of habitats. After cutting the increased light allows existing woodland floor vegetation to grow vigorously.

10 - Isabelle Duvivier. *Stillwater Cove*, 2019. (Image courtesy of artist)

9 - Transportation of logs through the valley, 2019. (Image courtesy of I. Mazzoleni)

Coppices ensure varied age structures and a diverse range of habitats, rich in the food and nectar sources on which creatures like butterflies feed. It provides a village residence with a variety of wood dimensions from a single tree.

As an architect, a person that single-mindedly builds shelters for humans, my experience at NAHR was like being trimmed down to the ground to resprout into many varying but related interests around the management of lifestyle, shelters, nature, and connectivity to everything around development. And I realized that in allowing myself to really express those varying new shoots, that I was stronger, more resilient, had more points of contact, and a more open-minded approach to learning about how people and nature work together.
Isabelle Duvivier, Architect

INTEGRATING TREES & DOMESTIC HABITATS

In my architectural practice, I am blessed to have clients that express their desire to be connected to nature. But often they think this means large windows framing views. Since the interest is there, I take the conversations to a deeper level of interconnectedness. This can be in the siting of the project to respect the flora and fauna of the place, the materials used, and how the building touches the earth.

My NAHR research looked at how vernacular architecture can be adapted to utilize building skins that function more like the tree bark of the region. I attempt to use this analytical methodology in my practice now when I am looking at building skin systems. I

11 - *Tremella*, jelly fungi, growing on a rotting branch in the forest floor, 2016. (Image courtesy of I. Mazzoleni)

try to view them as a filter, like skins in nature. These skins allow good elements inside while keeping bad elements out and allow the reciprocal from the interior.

Jay Yowell, Architect

Many people have an innate fear of nature, especially very open places where there is no refuge, and very enclosed dark spaces where one might get trapped. We fear many living creatures: spiders, slugs, snakes, maggots, lions, tigers, bears; things that move too fast, and things that move too slow. As an architect, I create environments that help transform fear into wonder: buildings that attract the kind of nature that we love to serve as a gateway back to biophilia. For example, humans love butterflies, which need both grasslands (for the wildflower nectar) and trees (for shelter) to survive. As a creative partner with nature, I design buildings with open areas that incorporate the celebration of the most insect-attracting beneficial trees, and shrubs or grasslands, so clients can enjoy all of nature (including butterflies and birds).

12 - Jay Yowell. *4 stages of Biomimicry and Architecture*, 2016. (Image courtesy of architect)

Two such tree/groundcovers are the coast live oak, which hosts plants for the larvae of 163 local butterflies and moths, and California buckwheat, which hosts plants for 51 butterflies and moths. The ginkgo tree by comparison hosts zero species. People that enjoy my creations recognize the beauty in the local, and embrace that nature is forever changing and can be messy. Reconnecting humans and non-human animals and respectfully co-existing with living organisms are the goals of our partnership.
Isabelle Duvivier, Architect

REDEFINING THE CLIMATE DISCOURSE TO INSPIRE FUTURE CHANGE

We can reframe our conversations by sharing knowledge in a creative, rather than abstract or judgmental, way. I view abstracting as a reduction to essential elements. It is not a literal interpretation, but a simplification of the image or idea. This is dangerous in our political environment (especially in the U.S.) which is so polarizing, that certain buzzwords (which have little meaning behind them) should be avoided. Showing the common ground, and finding connections that even opposing parties can see, will awaken change.
Jay Yowell, Architect

Through conversations, the challenges and issues discussed are more proximate. It is not only important to bring a critical perspective closer to our everyday lives, but also to operate on a level beyond words, informed by revealing artistic and creative processes. Hopefully through wandering, reformulating, or gaining awareness about how we inhabit a place or the kinds of relationships we establish with our surroundings through artistic production, it is possible to spread this practice to others through images in an accessible way. Art and creative processes are able to establish or point out connections that might otherwise go unnoticed.
Viviana González, Artist

We can inspire hope for the future through constant engagement. It is possible to be both deeply alarmed and concerned about the environment while also experiencing these spaces with wonder.
Caroline Clerc, Artist

A so-called 'good person' takes these sorts of questions as their lifelong ethical responsibility and suggests that this is the sort of question for which we should never expect a complete and final answer.
Dan Disney, Poet

You must take a stand for what you believe in and bring others along. Climate change is having an unparalleled impact on nature. Los Angeles is a global biodiversity hotspot with some of the richest natural ecosystems in the world, as well as the most threatened. In particular keystone species, oaks and Southern California black walnut trees, have proven to provide superior habitat and food for diverse forms of wildlife compared to exotic tree species. The city of LA has invested in a Biodiversity Initiative and has committed to no net loss of biodiversity. Every city department has been directed by the city council to incorporate biodiversity into their practices and report on their progress. Yet the current proportion of native trees in the city's public right-of-way is merely 3%, and even filling every vacant parkway tree well with native species would bring the overall native tree cover to 29%. The Community Forest Advisory Committee, CFAC, is requesting the City (of Los Angeles) commit to planting only native trees so that habitat & food can be provided to restore our biodiversity. This is not the way the city has ever managed its urban forest, and they are resistant. CFAC is taking a strong stand, taking some heat from the city, but we are not backing down. We are trying to change the dynamic. Don't shove trees where they fit in, put trees where they need to be and

change the infrastructure to accommodate the critical nature so it can thrive.
Isabelle Duvivier, Architect

proud of and that belongs, a land returned to the species that once occupied this place and were replaced by species imported.
Isabelle Duvivier, Architect

REFLECTIONS ON 'PLANTING'

If you have some dirt where you live, plants that are native connect you to the complex order of the dirt, insects, and climate of your environment.
Caroline Clerc, Artist

I am planting, in Venice, California, a coastal urban forest with native shrubs and trees that attract native birds and pollinators. But most importantly I am creating a collective vision of a future that all who get to know me/ us want to be a part of—a land that we are

I am trying to plant a way of relating to places by understanding them as a weaving web of relationships with which I am also woven. I believe that this is an alternative to try to reinvent/ remember how we inhabit places from relations of reciprocity.
Viviana González, Artist

Planting the smallest of inter-connected systems, the mycelium, because it is critical to the success and health of the woods and thus, our planet.
Jay Yowell, Architect

13 - Visiting the woods, the *Roccolo dei Salini* in the Taleggio Valley, 2021. (Image courtesy I. Mazzoleni)

14 - Elke Eichmann. *Imperfection*, 2016. (Image courtesy of artist)

Rock & Stone

Material Culture & Cultures of Making

CHAPTER 2

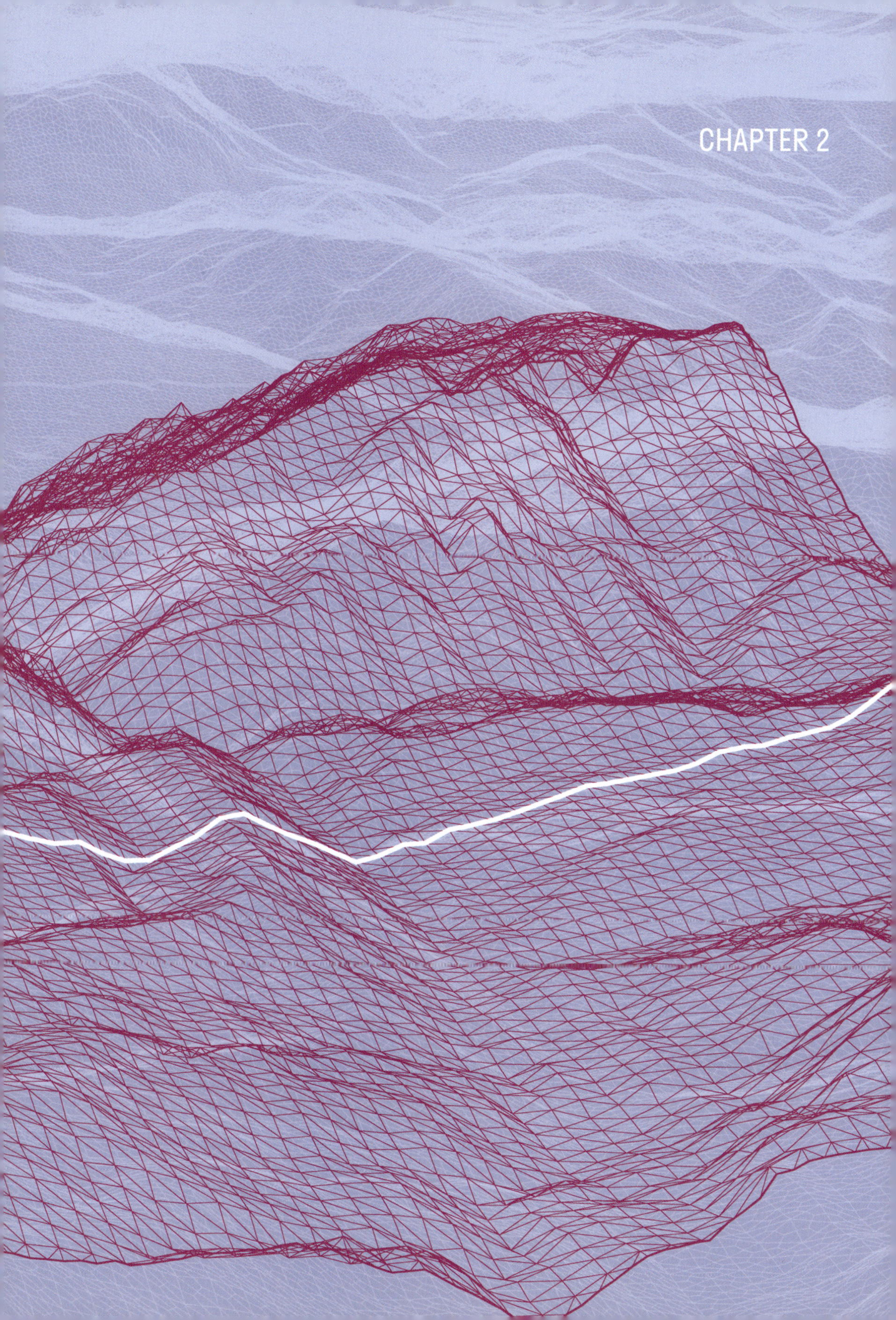

Introduction

Aslı Suner

Rock and stone make up the solid earth which has been providing the foundation for life on land to hold on to and evolve. As solid and permanent as this rocky earth layer seems from the point of view of modern human history, a fast-forward look at the Earth's 4.5 billion-year story shows the dynamic complexity of this foundation that is never really still. The motion of the crust is explained by the plate-tectonics mechanism which has given way to some of the amazing landscapes we came to inhabit such as the mighty mountain ranges like the Alps, as articulated in Stefania Cabassi's essay in this chapter. The mountain-building capacity of this highly complex mechanism coupled with the weathering and erosion of these very mountains has helped Earth to recycle its rich minerals from the depths of its mantle out to the surface and back to create some of the conditions for complex life to evolve. In the process, it revealed some of its internal layers which became the source of materials for our early tools and then the components of our richly varied human habitats. Softer layers were dug to make underground cities, and others were cut and transformed into marvels of art, architecture, and infrastructure. Eventually, with the adaptation of ever more powerful tools, deep layers of solid earth were extracted and processed through human design and engineering as they reincarnated in various forms as cities.

Geological layers may not be readily recognizable anymore in these complex urban settings, yet we are immersed in them. Our lives are entangled with the bedrock in multiple ways and scales from micro to macro; the iron in our blood or the stones in our kidneys are earth-borne. On the other hand at the macro scale; from the concrete that drives global urbanization, to the silicon that fuels the information age, we are in constant engagement with geology. Now, as more than half of the world's population is living in urbanized areas, the rate and scale of such engagements have grown immensely in parallel to rising cities and their expanding hinterlands. We dig large holes to excavate mineral resources, as we fill seas to create new landforms for further development. Consequently, the rocky layers of the Earth are transformed to such a degree through invasive processes driven by unsustainable growth economies, we humans find ourselves acting like a geological force, as you will read in Joshua G. Stein's thought-provoking essay in the next pages.

Consider taking two different walks and pay attention to the stones you might see. A walk by the beach of a pristine Mediterranean coastline reveals pebbles shimmering in light with changing colors acting like individual time capsules, each displaying their mineral signature through their beautiful and diverse patterns. These pebbles hold the clues of the mountains they once were, and hence the geological forces of the Earth. A second, similar walk by the seaside of a densely urbanized metropolitan area exposes pebbles of a different kind. These might be rather hardened and rounded-off conglomerates of rubble, that emerged from waste flows of nearby demolitions and constructions. These rubbles of concrete, porcelain, or terrazzo finishing materials, are components of urban transformation projects and the outcomes of human-processed geological layers. They are no longer bound to the deep time memory of the geological context of the place but are rather dispersed and highly processed aggregates from all around the world.

These two different encounters of course are direct reflections of the change in our material culture from pre-industrial times to the age of the Anthropocene. The close relationship of the craftsperson to the rock is both an outcome of the landscape and the driver of a culture of making that manifests itself in harmonious vernacular architecture as you will read in Stefania Cabassi's essay. In contrast, Joshua Stein's essay illustrates the radical shift in the culture as it is no longer an outcome of place, but rather of globally driven growth economies of extraction.

Moreover, contemporary material culture is mostly fed by the fast-paced consumption of objects and monuments, leaving the processes of making, and journeys of material transformation out of focus. For instance, we see the aesthetics of a marble floor for its beautiful color and patterns, and its valuable qualities for durability, rather than its epic journey across deep time and place from planetary bedrock to floors we walk on. Similarly, we build our ever-expanding cities with stone, concrete, or steel, and pave the connecting roads between our cities with asphalt which are all materials of geological origins. Yet we are hardly aware of the processes of their formation in our rather myopic contemporary material culture. Understanding their journeys requires a long-term view of planetary forces as well as an encompassing perspective into the far-reaching impacts of human-designed technologies. To face the consequences of our forces and making, the science of geology provides not only the context but also a powerful exploratory lens to understand our place in and our impact on Earth.

Geologist Marcia Bjornerud (2018) defines rocks as verbs, rather than nouns. Journeys of rock tell the stories of natural history entangled with the stories of human cultures of making. It matters immensely to recognize these interconnections as we are at a critical threshold paving the milestones of our collective journey into the future. This chapter attempts to reveal some of these journeys and stories through a geologist's perspective which takes a long view through time, an architect's outlook with a wide view across the planet, and finally, through multiple creative Fellows' voices using geology as a lens for provocative thought. The following offers reflections on many exploratory walks amongst the outcrops of Earth's ancient basin that rose into the mighty Alpine range in deep time.

The Picturesque Versus the Planetary

Joshua G. Stein

UNCOVERING BURIED STRATA

The stone and rock beneath our feet connect us to larger immense planetary formations that, although in constant flux over millennia due to thermodynamic and tectonic forces, we have perceived as largely static and permanent. However, we are becoming increasingly aware of yet another force and temporality at play that demands we reconceive this assumption; the planetary (trans)formations emerging due to human activities.

A summer hike in the Italian Prealps provides insight into the longstanding relationship between human habitation and the ground that supports it. The rock and stone of the mountain valleys offer access to a rich history of vernacular architectural traditions, as this area's craft and construction techniques emerged in response to the specific materials available. For example, their particular friability, strength, and ease of shaping determines the precise slope and patterning of the idiosyncratic stone roofs, connecting them directly to the bedrock on which they sit. This association between artifact and context creates an aesthetic that we understand to be harmonious with the mountains and valleys of the region. For a cosmopolitan traveler, this picturesque scene is a welcome relief from the cacophony of contemporary junk space that is instead generated by logics of profit and logistics.

But of course, it is impossible to keep the blinders fixed in exactly the right direction at all moments. Sooner or later, somewhere in the peripheral views of the region's dusky profiles and atmospheric perspectives, a particularly awkward cut into the mountainside reveals itself. Once this incongruity registers in the consciousness, a host of accouterments—earth-moving equipment, outbuildings with sifters and crushers, subsidiary and parent company logos—accompany and confirm the aesthetic blemish, dissolving the illusion of a purely bucolic landscape. But what at first appeared to be an anomaly is perhaps more accurately a glimpse of the larger economic substrate of these mountains. The region's underlying strata, both material and economic, extend the human-mineral interface beyond the harmonious aesthetics, connecting this particular valley to vast deposits, mineral and financial, that stretch across the globe. As it turns out, the logics of profit and logistics are not so deeply buried in these hillsides. What kind of worldview might we use to best comprehend the hidden layers beneath this scene? While one might be more drawn to the views of the picturesque mountains than the areas sacrificed to capital and profit, the vision that allows us to perceive with equal ease (and discomfort) the craggy cliffs as the network of gravel pits may be the one most necessary at this moment.

1 - The Cava di Arabescato Orobico marble quarry at Camerata Cornello, in the Orobic Prealps. Viewed from a distance (L) the quarry is barely noticeable within the larger picturesque landscape. From within the quarry itself (R), the logics of industry and commerce dominate, 2017. (Images courtesy of I.Mazzoleni and author)

Geology helps reveal the inner workings of the planet, at least its mineral aspects. While other disciplines offer a view into the physically microscopic or the remote, geology reveals that which is obscured by layers upon layers of history: processes which are only understandable as such after viewing them at immense scales, both physically and temporally. It is these very tendencies that allowed geologists to help recognize the depth of human impact on the environment. Until relatively recently, we imagined pollution, species loss, toxic spills, and the scarring of the landscapes as tragic, but discrete and episodic events. Even more naively, we saw most of these problems as somehow reversible: a contaminated dump could always be remediated. We could do and undo. However, the cumulative effects of climate breakdown have eroded this falsehood, making it clear that certain forces set in motion collectively cannot be undone by individuals, nor, even more terrifyingly, by the same collective. The reality that certain effects of climate change are already 'baked' into our future makes it clear that we are dealing with immensely large phenomena that imply an inertia that is difficult to fathom at a human scale.

The geological paradigm is an appropriate one through which to examine these forces and structures set in motion, or constructed, by humans, that bear consequences at a scale far vaster than our typical understanding. If the stone walls and roofs that inhabit the landscape are one version of human interaction with rock and stone, then the corresponding network of massive pits, quarries, and mines is another. However, these latter artifacts are not analyzed with the same criteria as our stone interactions at a smaller scale. It is hard to romanticize or aestheticize the landscapes of extraction. However, certain traditions within geology prioritize a material appreciation based on more than the purely aesthetic. The immense quantities of stone that have been subjected to incredible forces and transported vast distances, illustrate qualities that can be appreciated beyond a fascination with these superlatives. Instead, they reveal the inner mechanics of planetary transformation. The specializations within geology that connect material qualities to their economic potential can help us reveal the impact of the global metropolis on its material bedrock. How might we learn from the geological engineers who have mapped the underworld as a way to identify the precious resources necessary for our granite countertops, freeway infrastructure, and the literal foundations of almost every modern structure built over the last 100 years? How might we reshape our perception of the relationship between the stone pebbles and slabs that we build with and the transformation of our environment at a planetary scale?

PLANETARY DISPLACEMENT: BREMBANA VALLEY & SUN VALLEY

A geological focus on the relationship between the Earth's material and economic strata makes evident the ties between distant landscapes across the planet that might otherwise remain invisible. Two very different valleys: Brembana Valley, a sparsely populated semi-rural valley nestled in the Italian Orobie Prealps north of Bergamo, Italy, and also the Sun Valley, a flat, treeless, industrial district at the base of the Verdugo Mountains in Greater Los Angeles, California help clarify a new form of anthropogenically merged geographies. While these two valleys could not be more different from one another visually and culturally, they share both literal geological materials as two nodes in the same global supply chain, and typological similarities in terms of aggregate mining for the concrete industry.

The site of our earlier walk in the mountains, the picturesque Prealpine region of Italy that includes the Brembana Valley, is also the site of intense mineral exploitation which ties the geology to both the global marketplace and the large-scale terraforming of the planet. Part of this transformation is affected by the dimension stone industry that includes marble, slate, and other quarries of the region.

2 - The oversized presence of industrialized geology looms over a typical street in Sun Valley, 2023. (Image courtesy of author)

Dimension stone, as opposed to gravel, sand, or crushed stone, is natural stone that has been precisely quarried and cut to be suitable for architectural applications, both interior and exterior. These stone cubes and slabs are shipped to different markets across the globe, a certain portion of which arrive in Sun Valley, California.[1] Located fifteen miles northwest of downtown Los Angeles in the San Fernando Valley, Sun Valley is a more anonymous topography connected to Brembana Valley as a secondary stopover in the global supply chain of displaced natural stone.

While the forested vistas and green pastures of the Orobie Prealps make it easy to overlook the effects of industry hidden in plain sight, the lack of natural beauty in Sun Valley, lost in the industrial hinterlands of Los Angeles, makes it harder to ignore the impact of extractive operations on the mineral landscape. In examining the USGS Geologic Map of Los Angeles, specifically the San Fernando Valley that includes Sun Valley, we perceive what appears to be a pattern of seams of some kind of intrusive rock. In fact, these lines coincide with those of the 101, 118, 170, and 210 freeways, along with portions of Interstate 5. However linear these forms appear, the fact that they are indicated on a geological map asks us to consider them as something other than urban infrastructure alone.

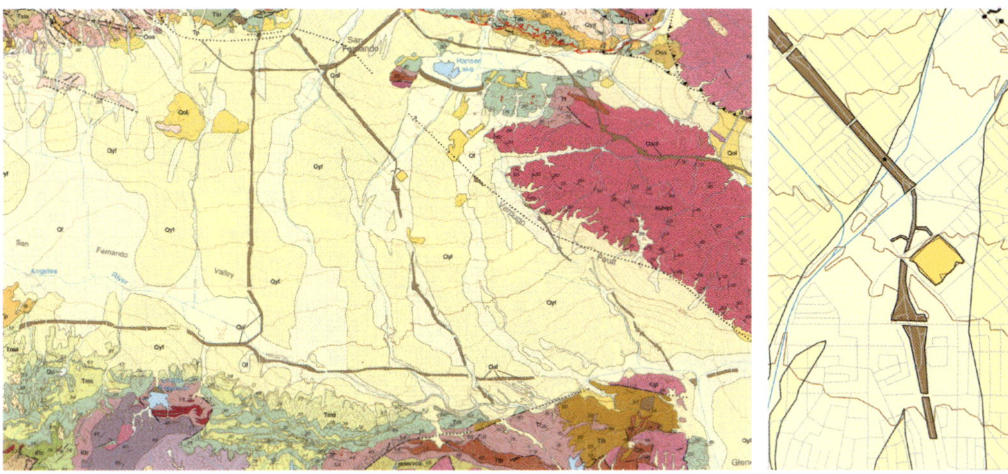

3 - USGS Geologic Map of the Los Angeles 30' x 60' Quadrangle, 2005, San Fernando Valley detail (L) I-5/SR-170 Freeway interchange in Sun Valley detail (R). (Maps courtesy of the U.S. Geological Survey)

Zooming in even further, we notice that these lines are not continuous and are instead interrupted at every half mile, perfectly aligned with the San Fernando Valley's grid of major thoroughfares. Here we see that this is a landscape formed entirely of constructed geology—engineered with a formal and dimensional precision that is more akin to that of architecture than landfill. What appear to be seam lines on the map are instead the embankments and fill of the elevated freeway

[1] Based on the availability of Orobico marble at one or more stone yards in and near Sun Valley (The Stone Industry 2022).

system, complete with the street crossings underneath that carve through this new geological upwell. These deposits are labeled as 'Qaf' or 'Quaternary—Artificial Fill', indicating that this geological material dates to the Quaternary Period and has been anthropogenically deposited.

4 - Overpass of the Interstate 5 Freeway in Sun Valley demonstrating the significant amount of fill used to elevate the freeway, 2023. (Image courtesy of author)

The glaring exclusion in this map, and most maps of this genre, is the omission of built architectural form. Although the massive volume of displaced materials contained within the region's single-family residences, commercial strip malls, and office high rises far outweighs that of the metropolis's freeway embankments, this 'architectural fill' seems invisible to the traditional geologist's surveying tools. While the geological map of Los Angeles indicates that Sun Valley is composed primarily of igneous granites and their metamorphic cousins like gneiss covered by sediments, if we were to take seriously the transformation of certain geological patterns induced by humans, we would also need to include in this map the recent intrusion of sedimentary rocks and their marbleized metamorphic cousins that have been displaced from the distant Italian quarries in the Prealps.

The term "glacial erratics" describes massive boulders displaced by the force of glaciers to landscapes that would otherwise have a very different geology. In Sun Valley we might identify a similar influx of foreign geology as 'supply chain erratics'. While glaciers displace geological artifacts over the course of years or even millennia, our particular supply chain erratics move at a much quicker pace as their journey across the Atlantic requires a matter of a few months. Conveyed 15,000 kilometers by container ship from an Italian port like Genoa, the Prealpine

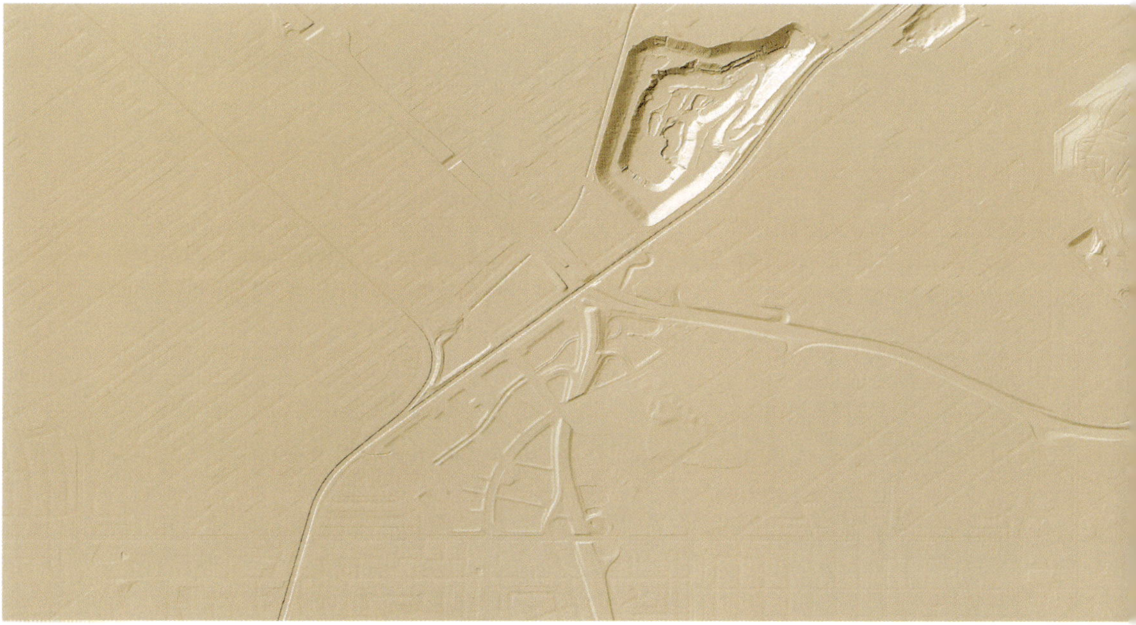

5 - A Digital Terrain Map (DTM) of the intersection of the I-5 and I-170 displays the anthropogenic mineral deposits of elevated freeways as well as the anthropogenic excavations in the form of quarries and below-grade freeway segments, 2023. (Image courtesy of the U. S. Geological Survey)

geological transplants slide through the Straits of Gibraltar, the Panama Canal, and are lifted onto the Pacific Plate for a momentary rest in the Port of Los Angeles. From here they are transferred to a semi-trailer truck or train to arrive in Sun Valley where they sit in stone yards awaiting their next transformation into residential countertops or boardroom wall panels. Sun Valley acts as a surrogate quarry—a resting spot for stone quarried from Italy and around the world and then shipped to the many local distributors serving greater Los Angeles.

LAVA URBANISM: SUPPLY CHAIN GEOLOGY

While glaciers offer one analogy for geological transport, the image of a volcanic eruption is perhaps a more apt metaphor for the planetary impact of mineral resource extraction as it more appropriately describes the speed with which the transformation is currently taking place. The output of the dimension stone industry and the concrete industry emanates from the Alps but also from the Apennines, which run from northwest to the south of the peninsula. Like lava slowly creeping down the side of a volcano, each year, millions of tons of rock are expelled from Italian quarries, snaking around the planet like a mineral flow inching its way along global supply chains and momentarily resting in distribution centers, depots, and stone yards. From there, these minerals winnow their way into urban and domestic fissures, insinuating themselves into kitchen countertops, sculpted facades, and donor walls.

6 - The most recent eruption of Mount Vesuvius in March 1944 discharged 10 million cubic meters of lava and ash across the landscape and atmosphere. (Image courtesy of Public Domain)

In 2018, Italy exported 0.8 million cubic meters of dimension stone, compared to 10 million cubic meters of lava and ash discharged in the most recent eruption of Mt. Vesuvius in 1944.[2] At this rate, it would take the stone industry 12.5 years to expel the same amount of geological material as that single eruption. Of course, Vesuvius erupts on average every 50 years, which means at current rates of export and eruption, Italy's dimension stone industry is quadrupling the output of the volcano. To complicate things further, the United States, the largest importer of Italy's marble, is simultaneously the largest exporter of marble to Italy. Italy imports an additional one third of its export tonnage from the US. But the US is not the only exporter of marble to Italy as Iran, Turkey, Namibia, and Brazil all supply Italy with significant quantities of marble (Stone News 2020). Due to a lack of regulation, marble is imported into Italy, cut or sculpted, and then exported as 'Italian marble'. The perceived preciousness of that status in many cases doubles the length of the circuitous routes this geology must follow, as marble may be quarried in one country, shipped to Italy for processing and then shipped back to its country of origin with a new name and new value added. With this increase in value comes an increase in carbon emissions.

While the supply chain movements of these materials appear to leave few visible marks along their ocean shipping routes, they manifest more obvious mineral traces along their earthbound journeys. The trajectories of this lava flow are facilitated by a network of parallel mineral pathways, as the transportation networks that move these materials across the lithosphere form new geologies at a vast scale. In Los Angeles we can see the mineralization of lines of international commerce,

[2] Calculated from data retrieved from Stone News 2019 and Earth Magazine 1944.

solidifying the logic of the global into the transformation of the planetary. The lines we traced on the geologic map of Los Angeles make clear how the transport of anthropogenic fill demands freeways built on embankments made of the same anthropogenic fill.[3]

Much of the mineral material for these seams of concrete flows directly out of the mountains nearby, especially the San Gabriels. Like the Prealpine regions of Italy, Sun Valley is a site of resource extraction for the production of concrete and asphalt. The Tujunga Wash creates an alluvial fan from the San Gabriel Mountains, producing aggregate that is then quarried in numerous pits. The Vulcan Sun Valley Rock and Asphalt plant is located along this wash, as well as Angelus Block, a major manufacturer of concrete blocks distributed nationally.

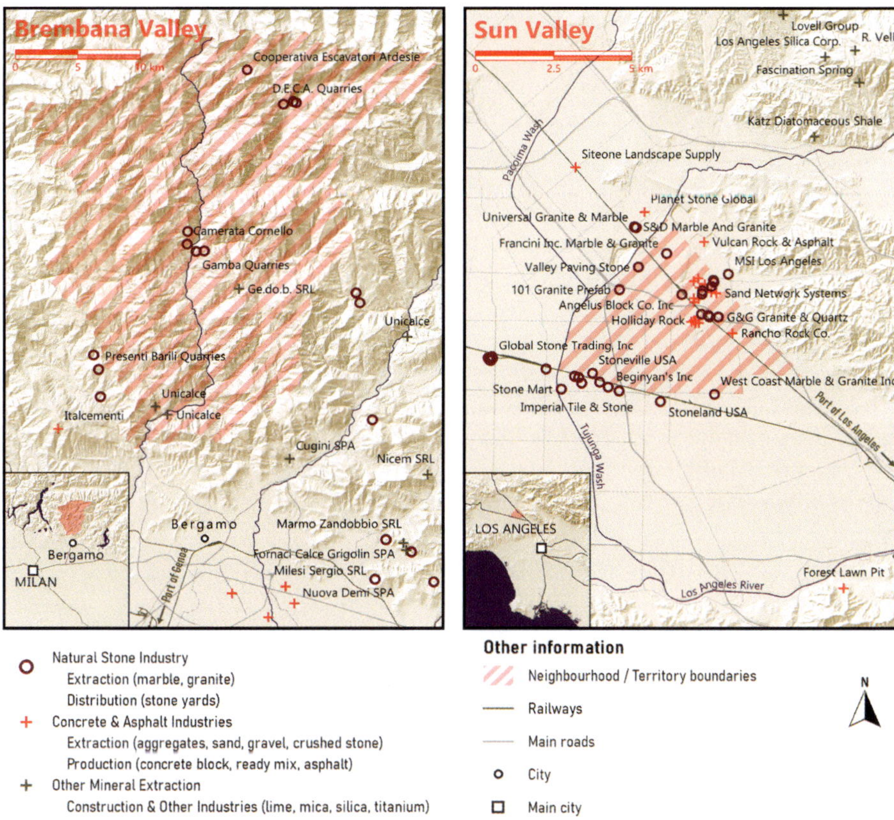

7 - Maps of extraction, production, and distribution sites in Italy and California reveal the direct flow of natural stone from the quarries of Italy's Brembana Valley (L) to the stone yards of Sun Valley (R). In addition to this planetary-scaled supply chain lava flow, each region straddles mineral outflows moving from nearby mountains to the aggregate mining and kiln facilities that transform these materials into airborne geology. Maps by the author and Andrea Galligari, 2023. (Image courtesy of author)

[3] For more on this concept, see Matthew Coolidge's research on the gravel pits of Irwindale, CA in 'Margins in Our Midst' in Varnelis Kazys, 2009.

While the connection between Brembana Valley and Sun Valley is most obvious in the shared literal materials of the dimension stone industry, the typological similarities continue through the anonymous concrete industry which is equally reshaping the planet.

As a supplier of the ubiquitous and banal material of urbanity, the concrete aggregate industry in Southern California and in the Prealps exists outside the general public's obsession with precious commodities like marble and beyond the lens of photojournalists like Edward Burtynsky, famous for documenting the extraction sublime of the quarries of Carrara. But this industry is ubiquitous, as the mere presence of mountains in proximity to urban areas is one of the few criteria for aggregate mining. These mountains and their alluvial fans provide a steady stream of geologic material ripe for exploitation by the concrete, construction, and asphalt industries. The banality of concrete here is key,[4] for the public has little interest in touring aggregate quarries. On the flip side of the supply chain, the public pays little heed to the provenance of the materials in the buried concrete footings of the house under construction next door, or the tilt-up warehouse flanking the freeways, or the asphalt and concrete surfaces that connect all these objects within the built environment. However, the aggregate industry slowly churns through the rock and stone of regions across the globe, transforming and redistributing one landscape into another.

ASH & ENTROPY: AIRBORNE GEOLOGY

In addition to the very literal movement of minerals across the planet's surface, these extractive industries challenge our understanding of geological materials as inherently earthbound. In fact, the most crucial impacts of industry on the planet are radically transforming the zone just beyond the earth's crust and the atmospheric impacts of the aggregate industry demonstrate that there is also an airborne version of geology that we must take seriously.

8 - Vulcan Materials Company's Sun Valley Asphalt facility viewed from the plant from Tujunga Av. (A) and detail of surge pile (B), 2023. (Images courtesy of author)

A few kilometers north of Bergamo, barely past the threshold between Padana Plain-the and the foothills of the Bergamasque Prealps, lies a lime production facility and limestone quarry of Unicalce, one of their 11 facilities in Italy that together produce 2 million tonnes of lime products annually

[4] For more on invisible, ubiquitous paving materials, see Laurence, 2019.

(Unicalce-Azienda 2021). The Unicalce quarry is a one-kilometer long cut into the face of a hillside. This is hardly at the scale of the quarries of Carrara, but neither does it compare in terms of the luxury status of marble. Instead this sacrifice of a portion of the landscape is surrendered to the more humble applications of construction, agriculture and other industries. The relative location of the Unicalce facility between the urbanized plains and the mountains is repeated the world over at similar thresholds. But these anonymous operations are slowly transforming the earth at a planetary, atmospheric scale.

Much of the limestone extracted from the Italian Prealps is destined for use in the production of either lime or cement. One of the primary detrimental effects of lime and cement production as well as aggregate mining for concrete is the particulate matter that is distributed across nearby communities. Dust clouds are visible at many of the mining facilities as quarried rock is pulverized and sorted across various conical mounds called surge piles. While this particulate matter might not cause heavy metal poisoning as would the dust clouds from some other kinds of mining facilities, it is a culprit in the chronic respiratory problems of local communities. Particulate matter less than 10μ in diameter, designated as 'PM10', lingering near these facilities is inhalable, and stays suspended in the air for long periods of time, increasing health risks. The fact that many mining operations are located in relatively close proximity to urban areas to save on hauling expenses means this impact on neighboring communities is even more intense. The residential communities adjacent to the Sun Valley aggregate facilities are listed in the top 5% of most polluted zip codes in California.[5]

Less visible than the dust clouds hugging the pits and quarries of the valleys is the 'geology' dispersed into the air as CO_2. In order to produce lime or cement, limestone must first be heated in a kiln to either 900° C or 1300° C respectively. This process, called calcination, necessarily releases CO_2, which quickly makes its way into the atmosphere. This means that a significant portion of our limestone geology is extracted simply to then burn into the sky. While the dust particles may hang in the valleys near the aggregate facilities, the CO_2 contributions from cement plants do not stay in the local environment, but instead mix with the airborne soup of other greenhouse gasses to evenly envelop the planet. And here we see the most direct link from smaller local quarries like Unicalce to the planetary scale: as with major volcanic eruptions, the plume of gaseous and airborne particulate matter from a single point source can drift around the entire planet. However, while major volcanic eruptions spew sulfur dioxide into the atmosphere, which has a net cooling effect on the planet, the carbon dioxide from cement production emitted into the atmosphere has a net warming effect as it is a greenhouse gas.

While the numbers are slightly lower for lime, for each ton of cement produced, almost one ton of CO_2 is released into the atmosphere (Clark 2021). This one-to-one relationship is a terrifyingly simple image. Although buildings are not made solely of cement, if we could imagine our cities as vast geological accumulations of minerals,

[5] In 2021, the census tracts in Sun Valley rank between 90 and 98 percentile for overall pollution burden. (California Office of Environmental Health Hazard Assessment 2023)

we can picture their airborne mineral doppelgangers floating in the atmosphere above.[6] While we imagine the release of liquid hydrocarbons from deep within the earth into atmospheric carbon as one of the primary contributors to a warming planet, the burning of solid minerals closer to the surface is also a key contributor. And of course, in addition to this phase change inherently releasing CO_2 into the atmosphere, the heating required to achieve the necessary temperature carries with it its own carbon footprint.

While we think of the making of the metropolis as a highly choreographed act of the careful assembly of materials, it is paradoxically the planet's most effective machine for liberating our geological materials from their distinct pockets and strata embedded in the earth's crust, entropically dispersing them into the atmosphere. The unwitting collective effects of hundreds of thousands of quarries and pits, crushers and kilns, each seemingly tied only to their local economy and landscape, is now redefining the boundary between lithosphere and atmosphere. At a planetary scale, this mineral upwelling and dispersal—thousands of 'volcanoes' spilling geology from the depths of the earth's crust across its surface while simultaneously atomizing carbon into the atmosphere—is the entropic flip side to the construction of the global metropolis.

9 - Aerial view of Midtown Manhattan, 2016. (Image courtesy of Suyash Dixit from Pixabay)

[6] Cement constitutes about one tenth of concrete and the carbon footprint for one tonne of concrete is about 72 kilograms (Clark 2021).
[7] Maarten Vanden Eynde performs a similar research expedition in his publication/project documenting the SD Worx building in Hasselt, Belgium. The publication *Building a Building* is dedicated to this research.

A GEOLOGICAL ATLAS OF THE BUILT METROPOLIS

Given the massive planetary transformations triggered by the construction of the contemporary metropolis and the commerce it fosters and demands, we must develop a method of both documenting this transformation and offering the public the tools to comprehend the scale of these changes. While the discipline of geology has an inherited predilection towards examining natural conditions untouched by human activity with an eye towards their exploitation, reversing this relationship could prove revelatory: how might we view anthropogenic modifications to the planet through the lens of the natural? If we could retrain the geological paradigm to instead examine all mineralogical processes equally, we might arrive at a method of surveying the planet that is appropriate for our moment. This Geological Atlas of the Built Metropolis reimagines the city as a continuous geological landscape, tracing material lines to suture the city together with sites and populations it has historically attempted to push outside its identity.

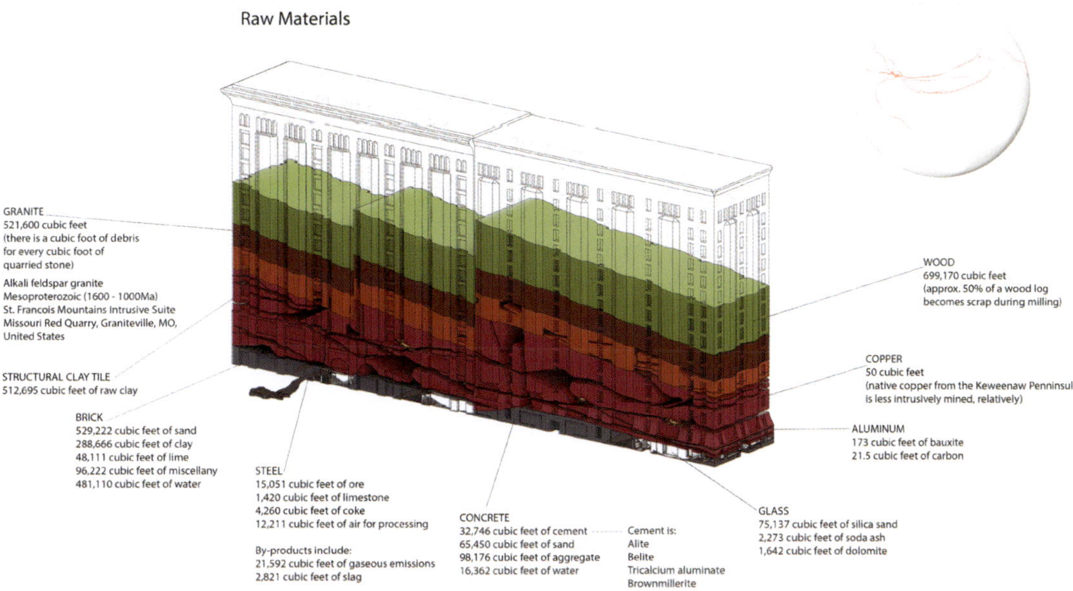

10 - Map of the planetary material trajectory of the construction materials for the Empire State Building in New York by Kiel Moe (R), 2017 (Image courtesy of K. Moe). The raw materials of Chicago's Monadnock Building represented as geological strata by Joshua G. Stein (L), 2022 (Image courtesy of author)

The act of Atlasing begins with simple documentation of the materials embodied in our built environment. A simple quantitative analysis offers a means of taking stock of the building stock of the metropolis, acknowledging the importance of counting in accountability. Architect Kiel Moe's studies into the origins of The Empire State Building and the Seagram Building in New York City both offer examples of what a detailed analysis of material quantities might look like. He categorizes these iconic modernist skyscrapers into their constituent materials, documenting the quantities of steel, stone, terra cotta, glass, rebar, and more—down to the tar paper used as weatherproofing.[7]

Beyond the static quantification of construction materials, we must develop a dynamic understanding of the totality of the materials that support the global metropolis as they move across the earth. Even more important in Moe's research is that he traces the journeys of these construction materials back to their origins, tabulating distances, mode of travel, and by extension their embodied energy. Combining both quantities and paths of travel, the Atlas acts as a geospatial database of the mineralogical and geological connections between the megalopolis and its hinterlands.

Landscape architect and theorist Jane Hutton proposes that we must include the sites of material origins as we examine our urban artifacts and environments. These distant and anonymous extraction sites drag with them repercussions on local environment and labor, becoming 'reciprocal landscapes', materially tethered to more familiar monuments, such as the Sands Quarry in Vinalhaven, Maine that provided the granite for New York's Central Park South Gate House. Hutton shifts the logic from one of commodities transported long distances to that of distant landscapes folded to meet and merge with one another. Here, the geologic tradition offers helpful metaphors, since geography is never static: as tectonic plates slide across the planet, today's snowy sedimentary rock formations were once yesterday's equatorial sea beds. Geological nomenclature and mapmaking strategies always acknowledge that matter is constantly on the move.

Finally, we must develop a sense of the geological implications of the anthropogenic movement of these materials as they scour, agglomerate, reshape, and redefine the planet. Our current image of global supply chains resides in the antiquated notion that these networks are somehow immaterial and ephemeral. Shifting to a geological paradigm admits that these global patterns carry planetary impacts. In order to foster a new public imaginary, we must also engage in a speculative act of representing this information according to new conventions. Imagine a meteorologist standing in front of a radar map tracking the movement of granite across the ocean from the Adriatic tectonic plate zone of Italy to the eastern edge of the Pacific Plate near Los Angeles. What other techniques must we invent? Could we imagine taking core samples of each structure in the city to document its precise mineral signature? Where would we find the heaviest concentrations of gypsum? In the interior wall partitions of global capital, or the stucco envelopes of suburban America?

By examining the extended metropolis and its architecture through the documentation of its constituent materials, their extraction, and disposal, we acknowledge some of the foundational injustices perpetrated by the metropolis against its marginalized hinterlands and their populations, however close or distant they may be from the urban core. We often imagine these sites of extraction to be far removed from the city itself, and many are in fact continents away from the place of use. However, some nodes within the archipelago of global supply chains are hidden in plain sight within the city itself. While they may register as terrain vague sites for some, they are neighborhoods or habitats for others.

If, as the term Anthropocene implies, we were to imagine the actions of humans as being a part of the geological forces, how might we reconceive of the various extractive industries in relationship to their host landscapes? What kinds of geological parallels can help us fathom the transformation these activities have wrought on the planet? If we take seriously the movement of minerals as they are implicated in the construction industries, we can trace their movement across a surprising array of landscapes.

CONCLUSION

While it is crucial to foster and maintain a sense of wonder about our ecological landscapes, our aesthetic predilection for the picturesque may not be the most trustworthy tendency at this moment. The techniques developed in the 'Geological Atlas of the Built Metropolis' intend to change the public perspective of our landscapes and shift them from a primarily picturesque experience to one we might call planetary. From this moment forward, we can no longer afford to indulge in purely aesthetic experiences. Instead we must process our policy and design deliberations through a filter that takes into account the cumulative effect of individual decisions and actions as they affect the material substrate of our existence. We must relentlessly connect incredibly local details to the immense scale of the planet, viewing each local instance as a manifestation of a larger tendency. Only then can we recognize what might be saved, restored, managed, reconstructed, or transplanted.

The tools and documents of geology have frequently been instrumentalized in service of the exploitation. Kathryn Yusoff's *A Billion Black Anthropocenes* examines how geology was, from its inception, born to classify and map territory in anticipation of exploitation-of resources, of human labor and dignity. Thus, the minute we begin to think beyond the stones and rocks of our immediate context and to strategically set about categorizing and locating their ease of access, we have transformed them into "resources" that lie inert, awaiting their ultimate exploitation and consumption. This legacy haunts the geological worldview.

However, on the other side of exploitation, in search of a post-extractive world, could these past sins become productive: the particular skills of the discipline somehow trained to undo some of the damage done? Could this tendency to envision the natural world as one made for human consumption, help to reveal the connections between any seemingly artificial artifact, and its site of extraction—its transition from specific material formation into generic fungible commodity?

This reverse engineering would develop a detailed biography for the materials recruited in the construction of the metropolis. The technologies associated with Building Information Modeling (BIM) already allow architects and engineers to ascribe data to each element involved in construction, from cost to embodied carbon.[8] Other

[8] BIM software packages allow architects and other construction professionals to create digital models as would Computer-Aided Drafting (CAD) software. However, in addition to modeling a building's geometry, BIM software embeds each building component with various data (manufacturer, precise dimensions, cost, weight, material, etc.) which then allows for a wide range of calculations (such as carbon footprint or estimated cost per square foot).

values could easily be added to this list, such as an index for conflict materials, the impacts on local labor and the environment. An even more advanced model would also offer a material history and narrative that builds a material memory by connecting landscapes across the planet and through time.

Remote sensing technologies—3D scanning and multi-spectral analysis—now allow us to dynamically track the transformation of our built environment and its slow return to the mineral conditions from which it grew. This demands that we adjust our ontology about the boundaries between the 'built' and 'natural' worlds. As we can more precisely follow the sources of our construction materials and the resting place of those same materials after they inhabit the form of buildings, we may have to revise our definition of architecture to include these states of prior- and post-building.

A new paradigm of geological thinking would extend the inherited geological perspective to include all mineral operations within the lithosphere (and perhaps atmosphere), whether we might at first deem them to be natural or not. The map of the trajectories of geological migrations from the mountains of Italy to the dense urban sprawl of Los Angeles is a first foray into the larger atlas that we must produce in order to train a new vision to discern a planetary-scaled landscape.

What Rocks Remember

Stefania Cabassi

Fathoming deep time is arguably geology's single greatest contribution to humanity. Just as the microscope and telescope extended our vision into spatial realms once too minuscule or too immense for us to see, geology provides a lens through which we can witness time in a way that transcends the limits of our human experiences.

—*Marcia Bjornerud*, 'Timefulness: How Thinking Like a Geologist Can Help Save the World'

KNOWING ROCK: GEO-LITERACY

Rock and stone are mostly considered inert elements indifferent to our contemporary lives. However, a closer look reveals a lost perspective from the realm of geological time that could inspire a paradigm shift over how we consider our existence on this planet. "Fathoming deep time" (Bjornerud 2018, p. 16) is critical in grasping the disruptive impact of human intervention in our collective habitat, and in designing new methods and tools to heal it in time.

The Taleggio Valley with its rocky outcrops is a place providing just this kind of perspective. It's a very small valley with less than 1000 inhabitants, located in northern Italy, downstream from the Alps, in the Prealpine region, a mountain chain that marks the border with France, Switzerland, and Austria. This valley is traversed

1 - Taleggio Valley *Orridi* Gorges. The rugged landscape is dominated by Dolomite peaks (L). Aerial photograph of the water cutting through a gorge (R), 2018. (Drone images courtesy of M. Wong)

by the foaming Enna stream which, in its evolutionary path, has carved a striking canyon (over 3 km) among dolomitic rocks with spires and towers that stand severely as watchtowers at the entrance of the Valley.

Winding along the canyon is the only connecting road, called the *Orridi*, between the main valley and the secondary: the construction of the road at the beginning of the twentieth century required the breaking of the dolomitic rock with the use of explosives, and the consequent preventative work to stop the looming and precariously balanced large boulders from falling onto the road below. Even today rock tumbles onto the road.

Once through the canyon, the view opens up showing decidedly less rugged morphologies, characterized by gentler reliefs of marly calcareous and pelitic rock outcrops and grassland areas dotted with typical stone constructions.

2 - Past the narrow gorges, the Valley opens to an expansive landscape rich with woods, grasslands, and local architecture in harmony with its surroundings, 2018. (Drone image courtesy of M. Wong)

This scene reveals the local architecture which is in total harmony with its landscape. Both a knowledge of rock and an understanding of its structure have been integral to developing the practices which gave way to a material culture that is well adapted in the context of the Taleggio Valley. Thus, the local architecture has emerged almost through a mimetic process alongside the close observation of these rock formations which have been shaping the valley over many thousands of years.

In a sense, one can claim that rocks have a lot to teach us if we know how to read them. Even a basic literacy in geology provokes a deeper understanding of our planet and our relationship to it over deep time.

GEOLOGICAL TIME AND THE ALPINE MEMORY

Mostly we humans understand history's timeline of events based on the length of our own lifetimes and experiences, which is limited to about a hundred years and not comparable with geological time. This short-term view has given way to an accelerated human intervention with geology through practices of extraction and terraforming, ignoring the long-term effects which consequently have played a crucial role in current environmental crises.

In comparison, the geological time unit is expressed in millions. It is the time elapsed from the origin of the Earth until today. Its duration (four and a half billion years) is dramatically larger than the timescale we understand in our human history. If we could scale down the entire history of the Earth to 24 hours, modern-day human societies would only appear at the last minute before midnight. On the other hand, rock tells a very long story, or rather preserves the space-time memory of several millions of years. That memory manifests itself in rock formations as traces of planetary history. In that sense, the rocks are like books that can be read to enrich our knowledge and open our minds to deeper and wider space-time dimensions. By following the traces left in rocks one can find indications, and keys, to interpret the present from the memory of the past with an eye toward the future.

In the specific case of the Prealpine memory, the rock outcroppings, whose feature of stratification from a layered deposition defines the region, had eroded from the surrounding lands, between 250 and 200 million years ago, into the sea basin which was located in the Taleggio Valley. This was a veritable sea that stretched across seven by sixteen miles and which for a certain period was paleo-geographically similar to the present-day Bahamas. Beginning 100 million years ago, and for more than 50 million years, the progressive closure of the basin and the concomitant formation of the Alpine chain led to the emergence of lands and the lithification and deformation of these sediments.

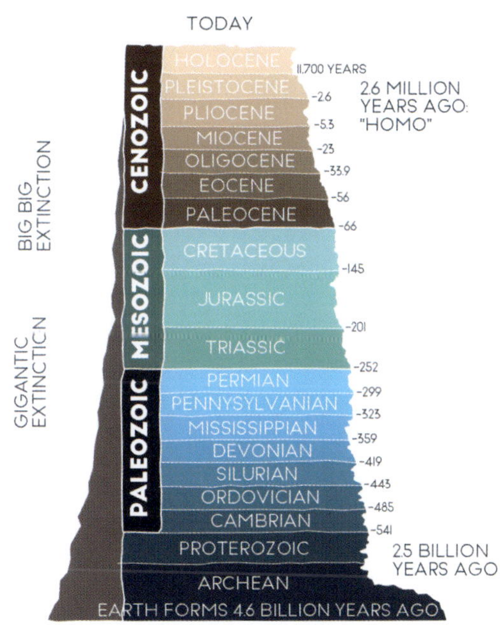

3 - Geological Time Scale. (Image courtesy of author, redrawn by NAHR)

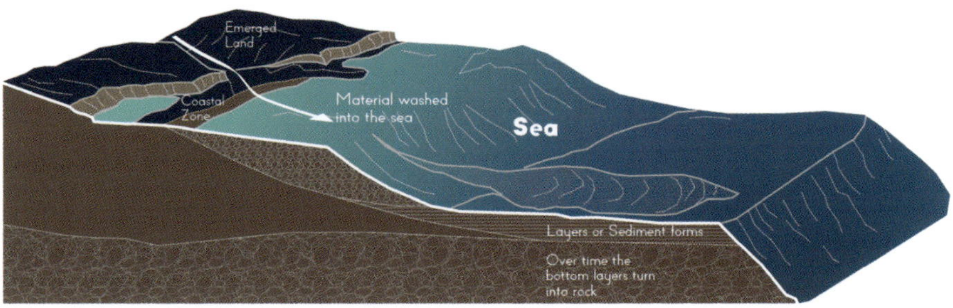

4 - Example of sedimentation in a marine basin. 2023. (Transect drawing redrawn by NAHR)

The geodynamic processes of plate tectonics are still evolving, although our perception of this seems limited within the human scale of space-time. Shifting our perception to the geological time frame is required to have the rocks reveal to us that their appearance and characteristics are impressions of memory traces from transformations that happened over millions of years.

In the stratifications of rocks and mountain ranges, we can find the oldest sediments formed by fossilized organisms deposited in layers that were initially horizontal. These sediments, just as in the formation of the Alpine chain, underwent deformations which can be explained by the movement of tectonic plates. The sediments were compressed, and their stratifications folded into the vertically layered rocks

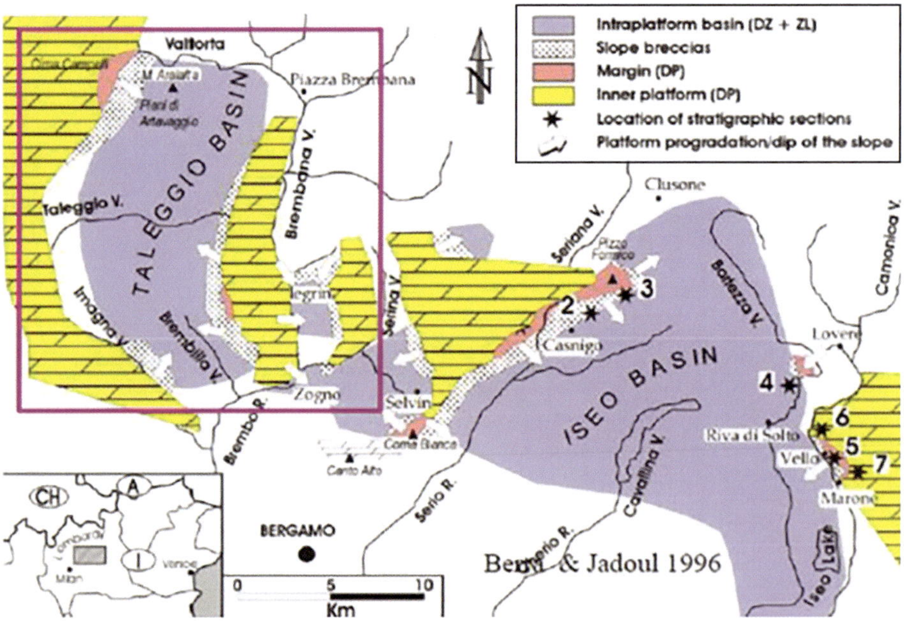

5 - Localization of the Taleggio Valley basin. A diagrammatic map showing the opening of the marine basin, and subsequent mountain formation due to plate closure. This determined the patterns of sedimentation and stratification of sedimentary rocks, as well as the lithological composition and structure in the region of the Taleggio basin. *Norian Serpulid and Microbial bioconstructions: implications for the platform evolution in the Lombardy Basin - Southern Alps, Italy - Model simplified by F. Berra and F. Jadoul*, 1996. (Image courtesy of Department of Earth Sciences Ardito Desio)

we see today. During this process, the rigid layers were fractured, as the plastic layers were folded according to the properties and arrangement of the mineralogical composition of the sediments.

The plates that form the Taleggio Valley are from both the African and European continental plates. Both plates were formerly united in the Permian, 225 million years ago, at the time of Pangaea (a term from the ancient Greek meaning 'the whole earth').

At the end of the Paleozoic, about 250 million years ago, the great continental mass stretched across the Earth's surface, surrounded by a single large ocean called the Pantalassa; the primordial Pacific Ocean. The Alpine chain that runs along the edges of the two plates is the trace of this tectonic encounter. The result is the mountainous structures and rocks that constitute the reliefs with their peculiar characteristics, which made them attractive as building materials, and subject to extraction for centuries, defining the material culture of the region.

THE MATERIAL CULTURE AND THE CULTURES OF MAKING

One of the local rock formations which provides materials in the valley is the Main Dolomite. It has a thickness of about 1000 meters and its composition is consistent throughout. These rocks were formed in the Mesozoic age, Upper Triassic, a time frame between 228 and 210 million years ago. They are mainly calcium carbonate enriched with magnesium. In this period the region of current-day northern Italy was in an intertropical zone, with warm seas, lots of light, and an arid climate. This was ideal for the deposition and synthesis of limestone by organisms like algae, bivalves, gastropods, and corals that over millions of years have been able to build organogenic reefs.[1] The fossils in these rocks are numerous. The invertebrates have only their inner form preserved, created following the filling of the cavities by sediments, which after their transformation into solid rock, reproduce an exact copy of the inner shell of the animal. The rocks are mostly gray or whitish gray, compact and massive, with indistinct stratification. Continuous subsidence of the seabed, a slow and gradual lowering, led to the formation of these carbonate sediments. The genesis of the Alpine chain, forced the coral reefs to emerge from the sea depths, exposing them to the conditions of exogenous agents such as rain and wind, which gave the region its unique characteristic appearance of pinnacles and white-grayish spires.

Another rock formation, Argillite of Riva di Solto, is also widespread in the Taleggio Valley. It has a thickness of about 700 meters. These rocks were formed in the Mesozoic age, Upper Triassic, a time frame between 215 and 204 million years ago. These are sedimentary rocks, formed through a progressive layering of limestone and clays. Limestones were formed directly in the sea basin, whereas the clays originate as eroded lands run-off onto the marine basin floor. The limestone found

[1] The very same type of reef formation can be found and studied in the Bahamas today.

in these layered rocks is used for construction. Throughout the paleogeographic evolution of the basin, the deposition of the sediments occurred in cycles of alternating mineral compositions due to the varying relationship between the basin's position and the emerging lands.

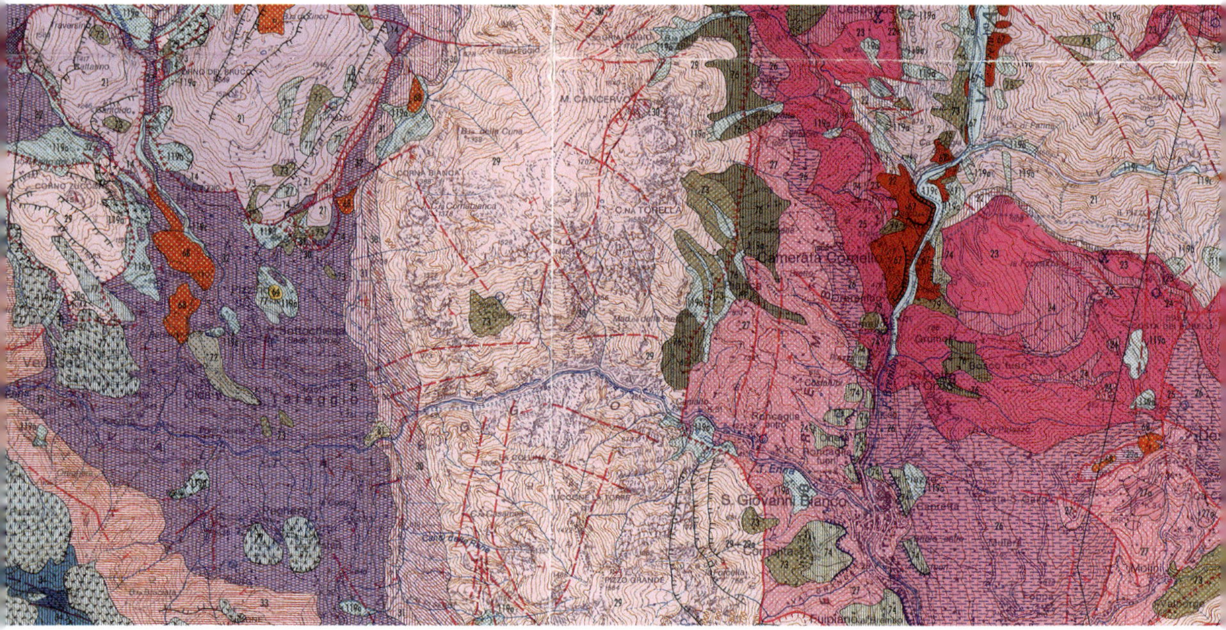

6 - Taleggio Valley map. In particular: in pale pink Main Dolomite rock, in violet Argillite of Riva di Solto. Source: Various Authors 2000, Carta Geologica *(Geological Map) della Provincia di Bergamo e Note illustrative* Servizio territorio della Provincia di Bergamo. Dipartimento di Scienze della Terra dell'Università degli Studi di Milano, Centro di Studio per la Geodinamica Alpina e Quaternaria del CNR. (Image courtesy of Dipartimento di Scienze della Terra dell'Università degli Studi di Milano, Centro di Studio per la Geodinamica Alpina e Quaternaria del CNR)

Consequently, today one can observe various stratifications within the Argillite of Riva di Solto that are cyclical and repeated, such as black argillites, black argillaceous marls, dark gray marly limestones, micritic limestones para conglomerates, and thin bioclastic intercalations. These different aspects or types of stone within the limestone are defined as *facies*, a Latin term meaning 'appearance'. It is a fine-grained limestone, with a homogeneous and compact texture. The final geological deformational phases of compression and folding gave the rock the typical *lastroid* (meaning centimeter-thick layering) hence the term, *piöda* (meaning smooth stone slab; *piöde* plural). It indicates the quarrying method which exploits the rock's 'fissility and cracking' systems. Thin layering is the result of strong deformation processes during the alpine chain formation on fine sediments that react to deformation in a plastic way, adapting to it by laminating themselves. This makes it possible for craftsmen to follow these laminations and extract materials without resorting to explosives.

7 - *Piöde* roof and walls typical of the Taleggio and Imagna Valleys. Detail of Ecomuseum's Baita Magrera (A), 2021. (Image courtesy I. Mazzoleni) Pesenti Barili quarry of Argillite of Riva di Solto in Berbenno. It is possible to appreciate the layered structure of the marly/pelitic rock formation (L). (Image courtesy of author) Roof in *piöde* (R), 2018. (Drone image courtesy of Val Brembana Web)

Guided by experience and careful observation of the specific compositions and appearance of the rocks, craftsmen were able to determine where and how to source their building materials. They could detect the cracks that constitute planes of structural weakness to extract stones with minimal effort. In this use of the materials there is an inherent respect; in Italian often the term 'cultivation' is used instead of 'extraction' suggesting a sense of harvesting rock and stone as though they were, too, living beings, like plants. Until the pre-war era, this level of care was given to the construction of buildings with these stones, reflecting on the trails of the deep time geological formation of the region, whilst building the anthropogenic layer in its continuum. Through this close and careful observation of the landscape dominated by the rocky outcrops, as well as an attentive engagement with the material, people of the region developed building practices, crafts, and techniques to use local stones for building their shelters, villages, and cities in harmony with the natural landscape. The local houses, with their dry stone walls and unique roofs, have appeared almost through a process of mimesis, as the emerging colors, patterns, and geometries often camouflage the architecture within the surrounding rocky landscape.

8 - Stone wall (A) and natural rock bank (B). Here there is almost no difference between the natural outcrops and the man-made wall, 2019. (Images courtesy of author)

An example of this harmonious connection with the local landscape can be seen in the process of selecting stones for construction; the skilled craftsperson strikes the stone with a hammer and listens to the sound emitted to figure out which use is the best. Like a doctor listening to the heartbeat of the patient with the stethoscope, the craftsperson listens. If the rock 'sounds good', it indicates greater mechanical strength and that the material is, therefore, suitable for the formation of pavements stressed by higher pressures. These pavements are built with stone cubes by craftspeople, who through their deep know-how create perfectly joined pavements, creating a continuous surface out of individual small parts. In a way, it is as if humans have learnt to pitch the rocks, sounding them out with a tuning fork, towards the aim of synergistic cooperation.

The farmhouse constructions called *baita* are unique to the region and their special roofs have been possible only because of the availability of a special stone from a particular formation in the area. The vernacular architecture of the Taleggio Valley is distinguished by the heavy roof in *piöde*, which is made with limestone slabs six to eight centimeters thick. These roofs were used to create both small stables for animal recoverings scattered throughout the pastures and larger houses for human habitation. Just as the ancient geological formations can be seen reflected in the construction of the vernacular dry walls, the traces of the prehistoric processes in the earth can also be appreciated in the shape and color of these steep black roofs.

9 - Understanding the structure, angles, and edges of the stones, to achieve the perfect fit. 'Stone carpet' making in Salzana, Workshop led by Architect Alessandro Mendini, Taleggio Valley. NAHR, 2017. (Images courtesy of L. Foresti and I. Mazzoleni)

10 - Farmhouse in Casinella, Pizzino, with traditional black roof and stone walls made with Argillite of Riva di Solto, framed by the Dolomite mountain formation behind, 2018. (Image courtesy of I. Mazzoleni)

The roof structure consists of large wooden trusses, composed of cylindrical beams, above which there is a smaller structure. Then there are the thick slabs of local stone, laid without mortar in overlapping rows. Because of their considerable weight, the stones are not laid parallel to the structure. They are placed horizontally on top of each other, requiring the use of a greater amount of material, so that most of the weight falls on the masonry. The ponderousness of the roof demands large walls to support it. The slope of the pitches can reach angles of 60 degrees. The buildings, therefore, tend to develop upwards in height and have simple rectangular plans. Thus, one can appreciate how geology has informed the local vernacular building morphology.

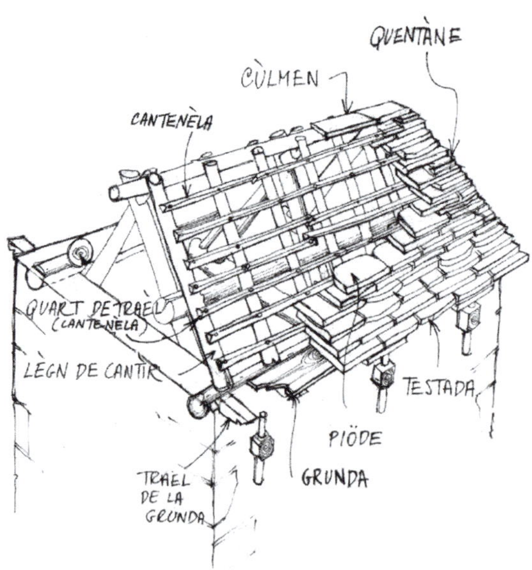

11 - The barn: vernacular black roof architecture in the Taleggio Valley. Drawing by Cinzia Invernizzi from the book "Prida e pioda", by Antonio Carminati and Pietro Invernizzi, Edizioni Centro Studi Valle Imagna, Bergamo, 2012.

MILESTONES

Through this kind of traditional environmental knowledge, the people of the valley knew how to coexist with the richness of the land. However, in the post-war period, the advent of tourism undoubtedly brought new demands and a period of intense building activity, in clear contrast to the typical stone-building practices of the area. The *piöde* roofs of houses still inhabited (at least in summer) are gradually being replaced by ordinary tiled roofs, both due to the scarcity of craftsmanship available and to the high cost of maintaining these traditional techniques and materials.

As the material culture evolved to adapt to the accelerated pace and demand that scaled up from the localized to the global, building and extraction practices became more aggressive and invasive in the environment. Supplies of the treasured stones that took nature millions of years to process and then expose, which in turn nourished the slow emergence of the local material culture and vernacular architecture, were exhausted in only a few hundred years. In the Prealpine region of the Taleggio Valley, a collective cultural perception emerged from the built heritage due to the characteristics of the rocks, allowing people to comprehend the links between culture, geological history, and the natural world, so we might live in harmony.

However, as the global profit-based interventions scaled up the extraction practices, a disconnect emerged between the material culture and nature. Moreover, these practices are also known to contribute largely to ongoing environmental crises. As the delicate balance between people and landscape dissolves, hand-crafted methods of slow extraction get replaced with fossil-fuel-based machinery. As pits grow larger, the surrounding ecosystems and related material cultures are disrupted too.

This brings us full circle to the urgent need for geo-literacy. It is no coincidence that the term 'milestone' means a guiding stone on a difficult path. Rocks tell a story of deep time that we must remember at this moment as we shift the paradigm and redesign our systems to be back in tune with all natural cycles, including the very slow ones. They preserve the memory and the trace of evolutionary processes that are cyclically repeated in time and space. In a sense, they are the milestones to help us decode the murky language of time, and practice long-term thinking to become better planetary citizens.

Voices of the Fellows
Rock & Stone

Roxanne Korda, Editor

The Alps, and the Taleggio Valley in particular, have provided a rich canvas of naturally occurring rock formations as well as human-made structures built of stone for exploration and artistic experimentation. These rocky mountains, soaring above historic *baita* farm buildings, with dry walls composed of looping geological and architectural forms, present a very beautiful and thought-provoking context; juxtaposed, at times, with the scarred landscapes of the nearby quarries excavated to provide building materials to the near and far cities.

The stuff of rock and stone has typically been embraced as hard, unyielding material. Yet a closer look reveals stories of slow-moving porous boulders worn by time, or of permeable and impermeable materials alike shaped by living organisms and natural elements. These shifting shapes and colors extend an intriguing invitation to engage with the unique material culture that has emerged in the Prealpine landscape of Northern Italy.

The following nuanced observations of rock and stone elicit an engagement with a familiar element, as unfamiliar and enigmatic interpretations.

1 - April Greiman. *Dossena Mine*, 2017. (Image courtesy of artist)

2 - Pietro Invernizzi sculpting. Jill Anholt. *Acqua: Vita*, 2018. (Images courtesy of artist)

ON MATERIAL CULTURE AND CULTURES OF MAKING

The culture of making and craft using rock and stone is embedded in our evolution as humans. Ubiquitous in the world around us, rocks and stones have served as and in shelters, structures, tools, and of course, sculptures, photography, and art. Perhaps our intuitive desire to feel the material, pick it up, and engage with its colors, textures, and veins, is embedded in a history of humans fascinated by the durability and malleability of stone. Inspired by the natural forces that can shape it, geomorphology, and geologic carving, compressing, and sculpting, we have learned how to manipulate rocks and stones purely by observing the forces nature projects upon the material.
Greta Ruedisueli, **Architect**

The long-lasting quality of products that can be made from them is a great incentive to create functional and beautiful objects with these materials, such as buildings made from stone blocks and carved stone sculptures. This same strength and durability make it difficult to manipulate rock and stone though, meaning the methods of making these products with them must be developed carefully over time. This development has thus evolved into a culture of material and making in several cultures and industries that have survived over centuries.
Katy Bradford, Structural Engineer

Stones moved into patterns by humans for the purpose of beauty or orientation like the ancient stone circles at Stonehenge or Avebury in the UK as well as near Evora and at Tamera in Portugal have been an inspira-

3 - Stonehenge, Salisbury, UK, 2023. (Image courtesy of I. Mazzoleni)

tion for building a small stone circle The stones were broken slate tiles that were discarded in the corner of the garden. This was a massive scaling down of my idea to create a stone circle. I originally wanted it to be way up in the mountains, but I didn't have any machinery, collaborators, or time. So, I made a small garden and held a ritual in it during the transit of the full Blood Moon. The story was then, I guess, one of grifting the stone, rearranging it, and creating a surprising effect not through monumentality, but through humility.
Simon Eales, Poet

5 - Kim Kwangwoo. *Stone Monument*, 2017. (Image courtesy of I.Mazzoleni)

4 - Alice Guarisco. *Stone light*, 2016. (Image courtesy of L. Agnelli)

Currently, it seems like the relevance of new technology in material cultures lies within the replacement of human labor. Robots can assemble stud walls and install windows, much like a human would. Pottery can be 3D-printed, and sculptures can be carved using programmed CNC routers. In all these examples, the technology has been developed to simplify the fabrication methods used for existing materials, many of which have been used for centuries. My current research utilizes digital fabrication methods to construct with natural materials. My main focus is 3D printing with natural fiber-reinforced matrices of clay, stabilized earth, and geopolymer materials.

The use of technology and software, such as modelling and analysis software, programming and robotics, and characterization of samples, allows me to be more efficient in verifying the performance of sustainable materials and fabrication methods. I do hope that the materials and methods I develop can be utilized in more simple ways though, without the use of any technology, after they have been vetted. As we continue to develop these various fabrication technologies, I think new materials will arise that are created for the technology, rather than the other way around. This could create whole new material cultures that are completely dependent on the technology that inspired them.
Katy Bradford, Structural Engineer

6 - Frank Melendez, bioMATTERS. *Meso Column*, 2023. (Image courtesy of artist)

7 - Katy Bradford. *New Technologies in Material Culture*, 2023. (Images courtesy of engineer)

ON THE ROLE OF TIME

The most important thing to think about here is slowness. Neither rock nor stone is inanimate, and neither is dead. It is a material that is very, very slowly, and internally, moving. This is not to mention that other, vital living beings depend on it: lichen and moss, for example. To provoke with rock? We can throw a small rock, of course. The history of human engagement with this material is mixed. But thinking about rock—to take it as a provocation for a kind of culture—we must no longer deduce feelings of stasis and solidity, but an incredible incrementality of movement. This kind of culture is potentially healthier than ones idealising rapid adaptation, conservation, or return.
Simon Eales, Poet

In the Taleggio Valley, I found myself fascinated by the lichen and vegetation growing through and eroding away the architecture of the valley. Quarries and the craft of construction are legible in the stone walls, roofs, paths, and structures throughout the valley. The composition and layout of stone is the initial road map for the movement of water, soil, and plants through a wall. Over time, stone structures increasingly yield to mounded, topographic forms of grasses and mosses, with traces of the architecture slowly disappearing into the fields and forests.
Greta Ruedisueli, Architect

8 - This image inspires the idea of Incrementality of Movement, 2017. (Photography courtesy of L. Foresti)

9 - Moss and lichen create a miniature landscape in a stone wall in the Taleggio Valley. Greta Ruedisueli, 2017. (Image courtesy of artist)

10 - Ho Ting Wei. *Who's Tone*, 2017. (Video still courtesy of artist)

ON 'JUST BE A STONE'

Who's Tone tells an unusual conversation between the character and different stones. The conversation provokes a sense of deep time and a contemplation of self as part of it.
Ho Ting Wei, Animator

'Just be a stone' makes me consider the vast scale of individual humans within time and the universe. In the face of climate change and disaster, we could learn to be more like stones: fluid with our environments, accepting of change, and no longer resistant to natural flows and forces.
Greta Ruedisueli, Architect

Today, we humans are continuously in motion, constantly using energy and resources. For instance, the most consumed material on earth, second only to water, is concrete, which requires a great deal of energy and heat to produce from mined, finite resources. Rocks, on the other hand, are formed over long periods of time in their given location by slowly flowing ions underground. They develop a comparable amount of strength by taking their time, rather than using large amounts of energy. In terms of sustainability, rock and stone can teach us that sitting still can build our strength, simply by using what we have around us. In fact, one of the best ways for us to mitigate climate change is to use less and work with more of what we have locally.
Katy Bradford, Structural Engineer

In my body, this comment registers as permission to relax very deeply. To release my jaw muscles and diaphragm, to let my eyes glaze over a little, and to let the drama of what has happened in my day recede from my awareness. I feel the heaviness of the stone in myself. I feel no unease, as one might expect, associated with freezing. I'm not concerned that the suggestion might be that I should be motionless. I feel even more connected to my humanness than I did before reading the comment. Stone emphasizes interconnection. I feel extremely interconnected, like a stone is. What I would say back is: 'I'm falling into myself'.
Simon Eales, Poet

ON ROCK POETRY

Can we sit in the face of something we don't understand and just let ourselves create meaning from that experience? Or not at all: just experience it?
Simon Eales, Poet

11 - Poetry resting upon its seat, 2017. (Image courtesy of S. Eales)

I would characterize the poetry I made when experiencing rock like this as in awe of rock, and inadequate in the face of it. I felt that the only reasonable thing to do was point out what rock seemed to be doing, find nice places to look at it, and make inarticulate sounds that attempted to convey my submission to it. You could follow my poetic descriptions of where ruined *baitas* were. I am still obsessed with this phenomenon, of reading something and trying to interpret it accurately. It's difficult enough with its instructions on how to build Ikea furniture, but it's basically impossible to do if you're reading poetry. And I love that. Reading and interpreting poetry is poetry too or requires poetry. One of the things I was doing with my *baita* project was creating a 'lite' version of that, where you had to see if you could interpret the poetic direction with enough accuracy to make it to the ruin. I had set up a little stone seat, and placed a poem, in each spot, as a reward.
Simon Eales, Poet

prickly rocks. desirous to see the end
within limits or find the
under thing he attempted
to explain, you trepidate
your way upwards,
navigating
wetness
rockiness
sticks and prickliness
zen yes smile yes sit and

give away

12 - Simon Eales. *Prickly Rocks*, 2017. (Poem courtesy of poet)

13 - Anne Elvey. *Shard, above Sottochiesa cemetery*, 2017. (Image courtesy of poet)

Yes: we can sit with something we do not understand and create meaning from it. Also, I believe, we can sit and forget meaning just experience stars above us? Science informs us of their origin, as of our own, we are no longer dependent on the old stories on creation, yet can we really, on a deeper level comprehend this infinitesimally small chance of existence? As a Finnish poet said: "We must embrace even life's long hours of illness as the short and rare moments when the desert is in bloom" (Translated quote from Edith Södergran).
Sofia Malone, Artist

MADE OF STONES

Geologists have a saying - Rocks remember.
—*Neil Armstrong*

14 - Juan Barte. *Material culture, vernacular stone architecture without architects*, 2017. (Images courtesy of artist)

15 - Imponent open air quarries in the Brembana Valley, Bergamo, Italy, 2017. (Clockwise image courtesy of J. Barte, H. Wei, G. Pellegrini)

Water

Vital Flux, Energy in Transformation

CHAPTER 3

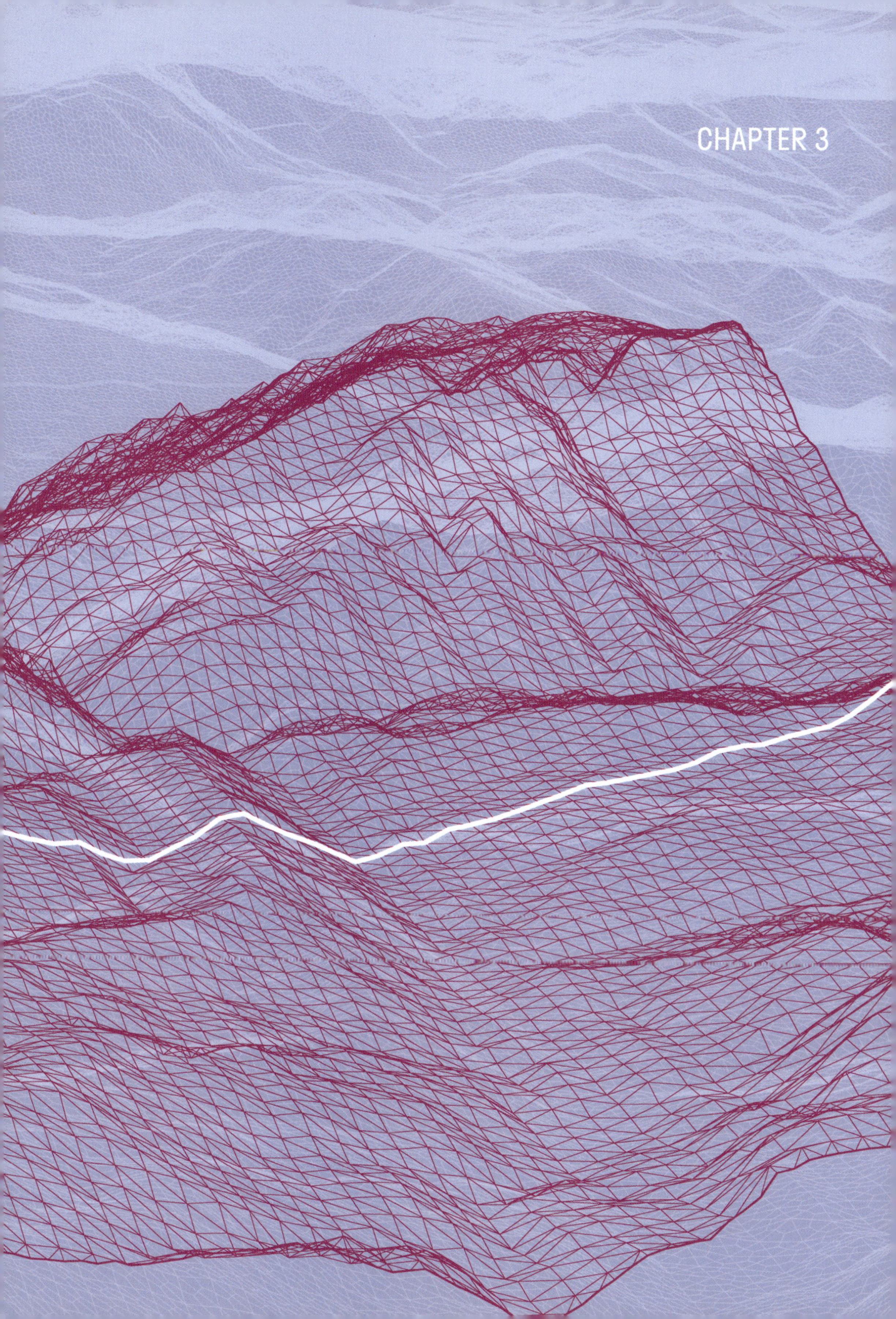

Introduction

Deborah Weintraub

Water is essential to all life. It is also a transformative source of energy that is stored in various states in the atmosphere, on the surface of the earth, and below the surface. It is the condition and qualities of local water that enable specific flora and fauna to thrive, and that characterize the ecosystem and the patterns of inhabitation. By weight it is the largest component of the human body.

A close reading of the earth reveals the manipulation of landforms by humans for millennia with the intent of directing, capturing, storing, using and dispensing of water, and extracting the stored energy in water flows and water gasses to operate equipment, whether directly or through the generation of electricity. This is true in all climate types, and for all ecologies. The symbiotic relationship between human inhabitation and water is a constant.

Using water as the origin of investigation and dialogue inevitably links to a multiplicity of other topics that includes social structures, governance, coexistence with nature, animal and plant characteristics, the form of human settlements, economies, spirituality, and more. In fact, the form of water infrastructures, with thoughtful analysis, can reveal basic values of a civilization. It is a lens to core social philosophies and to the structures of knowledge and beliefs, and to the relationship to the environment. In the arid western United States, for example, the extensive infrastructure of dams and reservoirs that have enabled intensive human inhabitation in urban centers and land cultivation in areas not historically cultivated, is a largely hidden phenomena. However, in recent years, with extended periods of drought and then a year of unexpected abundant rainfall, the politics and legal systems that govern this infrastructure are being publicly examined by a wider audience. The dialogues that have resulted have revealed anachronistic priorities born from prior times, and have made water policy a topic of broader discourse grounded in current social priorities. This is one example of how the relationship to water is a more primary indicator of human values than the relationship to other elements.

Current climate change, overuse, and the alterations to water cycles may have parallels to past events such as instances of species extinction and the demise of once robust civilizations. This is one foundation of a deep current fear that we may not be capable of understanding or addressing what we are witnessing in water cycle changes today, or capable of altering the global carbon-based economy and its historic negative impacts on water. The challenge is local for many—immediate and personal—but in our networked existence, it is a complex global discussion.

Water systems are under stress and the impacts will be felt on the natural environment and on human settlements. Sea level rise will remake coastal communities worldwide. Sea temperature will change the life that thrives in the ocean. Rain, floods, and surface ice melting will force settlement retreats and migration, and will restructure the characteristics of the patterns of plants and animals. Pollution and water quality degradation will change the landscape.

In the last two hundred years, with rapid industrialization and carbon-based societal frameworks, water sources have been mistreated. The perspective has been one of instrumentality with the core approach based on the question of how the manipulation of water could most efficiently support human settlements and industrial production, with little consideration of impacts. It is this instrumentality that we are being forced to reconsider.

If one agrees that all water is recycled water at a fundamental level, then all water use, systems of capture and transport, and systems of treatment and disposal are environmental questions. This perspective, whether it comes from responses to water-initiated disasters, responses to droughts, or responses to water contamination, shifts the thinking from economic metrics to ecological metrics. And in the ecological paradigm, the birds, the worms, the fungi, the foxes—they are all part of the discussion—as are the humans.

In the following essays, a variety of voices present perspectives on current water challenges at various scales, and from varied disciplines. The authors share an environmental perspective and the humility that is born from seeing human existence as part of the larger web of life. In the first essay the author focuses on northern Italy and its particular water regimen. It explores the transect of the high-altitude mountains—the Alps—as a water source, and the downstream challenges of water manipulation and precarity. It explores the science of recent changes to water stored as ice and snow at high elevations, largely the result of climate change, with the political responses to water crises felt downstream of these alpine sources.

The more global perspective of the second essay provides a historical perspective on harnessing water that has led to the hyper-industrialization of water with its now visible consequences, offering strategies for reconsidering the current dominant view of water engineering.

These two essays are followed by the more poetic reflections of diverse "voices" which personalize the individual connection to water in response to the landscape of the Val Taleggio.

As essential as water is to all life on earth, it is incumbent on all to make the current challenges transparent and central to personal, artistic, political and social dialogues. The impact humankind has had on water systems and the negative consequences of cavalier instrumentalism demands this conversation.

Troubled Waters:
Looking Forward with an Eye in the Rearview Mirror

Bruno De Meulder and Kelly Shannon

The world is at a critical tipping point. As politicians, decision makers, public authorities, business leaders and civil society largely fail to unite in setting and adhering to policies which tangibly and quickly address the adaptation to and curbing of global warming, the economic growth-at-all-costs paradigm thrives at national, regional (such as the EU), international and business levels. Not only is there a "code red for humanity," as stated by UN Secretary-General António Guterres (Guterres 2021), but also an existential loss of biodiversity, propelling what has been dubbed the "sixth mass extinction" (Kolbert 2014). The increasing frequency and severity of wildfires, floods, droughts, extreme temperatures and severe storms is rendering large swaths of the earth uninhabitable—by humans and non-human species alike. According to the *2020 Ecological Threat Register*, issued by the international think-tank Institute for Economics and Peace (IEP) and which compiles data to measure resilience and analyze risk in relation to extreme ecological shocks, the world has witnessed a tenfold increase in the annual number of natural disasters since the 1960s. The paper noticed that natural resources such as land and water are the most vulnerable. Nevertheless, there might still be a sliver of opportunity to respond to the dire realities of ever-increasing extreme weather events and climate-related disasters—change out of pure necessity. The radical reconceptualization of humankind's relation to both land and water is possible—and indeed a necessary condition.

Historically, a great number of settlements and productive landscapes were by necessity strategically located in relation to natural resources and traditional ecological knowledge (TEK). They largely developed in relation to a "knowledge base acquired by indigenous and local peoples over many hundreds of years through direct contact with the environment" (Inglis 1993, p. vi), and which were, at the end of the day and after trial-and-error, generally not over-exploited. As industrialization and urbanization accelerated, scientific ecological knowledge (SEK) supplanted TEK. Scientific forestry, modernization and industrial-scale agriculture and resource extraction, highly engineered water management, and land-consumptive urbanization radically and irrevocably transformed "the face of the earth" (Thomas 1956). Water urbanism was coined as a notion in 2008 (Shannon et al. 2008). The concept acknowledges that whether in the past, present or future, human settlement has an inextricable link to water. Settlement implies water management. Traditionally, settlements were strategically located in relation to coasts, lakes, streams, rivers, and the whole set of tree-like tributary structures including streams and creeks, with their corresponding

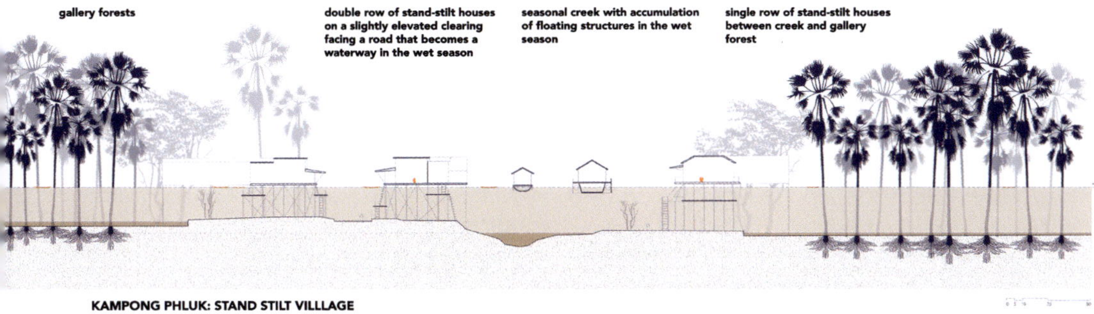

1 - The Kampong Phluk Village on Cambodia's Tonle Sap is designed to live with the rhythms of water flows, 2022. (Image courtesy of Research, Urbanism and Architecture)

profits and perils. In arid climates, and geographies not immediately adjacent to visible water sources, settlements were developed in relation to highly refined engineering techniques. A precision with regards to placement of settlements was necessitated to exploit, but not deplete, natural resources. The self-replenishing aspect of water resources was intimately understood, and extraction techniques were locationally nuanced to maintain the renewable facet. Today, in contexts and languages throughout the world, ancient settlement place names attest to the foundational importance of water. Settlements were often found at the confluence of water streams, along major rivers, on seacoasts and in inland and coastal deltas.

2 -Schelde Basin's cities and villages in Belgium are distributed along coasts and rivers (every 20 km) and tributaries (villages). Christian Nolf, 2013. (Image courtesy of C. Nolf)

Humankind has invented ingenious methods to catch rainwater, seasonally store and release water in ponds, tanks, basins and reservoirs, tap into aquifers at foothills to construct incredible irrigation systems, precisely divert water for irrigation from lakes and rivers, drain water from waterlogged terrains, reclaim land from seasonal or tidal flood plains in deltas or from the sea (by 10th century onwards, Chinese *wei* in the Yangtze River Delta no doubt inspired the 12th century development of polders in The Netherlands), and cleanse and convey it for daily use, both providing potable water and removing wastewater. Traditionally, water infrastructures largely pre-define the structure and fabric of settlements that anchor to it. In numerous, if not all contexts, water infrastructure was not only the primary physical armature, but also part and parcel of the symbolic and cultural ordering of the territory. The pragmatic and cosmologic were flipsides of one another. Numerous ancient examples reveal an intelligence in the economic and 'smart' combination of hard and soft engineering, hydraulic logics and socio-cultural beliefs and material practices. The land-based Khmer Empire (802-1432 CE) covering 20,000 hectares (49,421 acres) in north-western Cambodia, was developed as a bold system of water management which networked rainwater basins from the scale of the entire empire (with large *barays*, 1.8 by 7.5 kilometers/ 1.1 by 4.7 miles) to the individual household (with *trepang*, small ponds used for drinking water, bathing, and watering animals). Between these scales were a fine-mazed system of moats, channels, ditches together with dikes and elevated earthen platforms for both effective paddy irrigation and Hindu-Buddhist worship of the 'god-king' by way of 'temple mountains' (Pottier 2008).

A counter to the highly visible Khmer water system imprint is the ingenious 550-331 BCE underground water transfer supply system developed by the Persians on the slopes of piedmont alluvial fans (Cressy 1958). The gravity-fed horizontal conduits, known as 'qanats', and with lengths from 2-80 kilometers and a slope ratio 1:1000-1500, provided fresh drinking water and irrigated oases in the desert; they were constructed deep enough in the earth to pierce the underground water table and penetrate aquifers. There is a vertical 'chain of wells' which are necessary to oxygenate the water flow and for maintenance of the conduit, subtly visible on the surface of the earth. The importance of the qanats in both practical and spiritual aspects was evident in the fact that the qanats terminated at the ground level in bazaars, mosques and public or private summer living rooms, as well as by their accompanying invention of water clocks (one of the world's oldest time-measuring instruments, which often in the form of bowls, uses the flow of water to measure time). The engineering of qanats required a breakthrough in the development of mathematics, which can make one wonder whether to categorize it as TEK or SEK. It is in any case an archetype of water engineering that was fundamental for the genesis of the earliest cities in the Middle East.

In 1000 BCE, the Singhalese developed an ingenious system of tanks and reservoirs which were linked to irrigation and both sacred and profane settlement structures (Shannon & Manawadu 2007). Rome's Catholic church-driven urban transformation between 1560 and 1630 utilized both highly visible (aqueducts, fountains, bridges) and invisible (conduits, distribution tanks, sewers) water infrastructure to develop the city, and a system of public spaces reflect this infrastructure and remain a strong part of the city's contemporary identity (Rinne,

2011). In short, from a historical perspective, water structures are intrinsically intertwined with the built environment. Settling implies *settling with* the water issue, including pollution and other ecological issues.

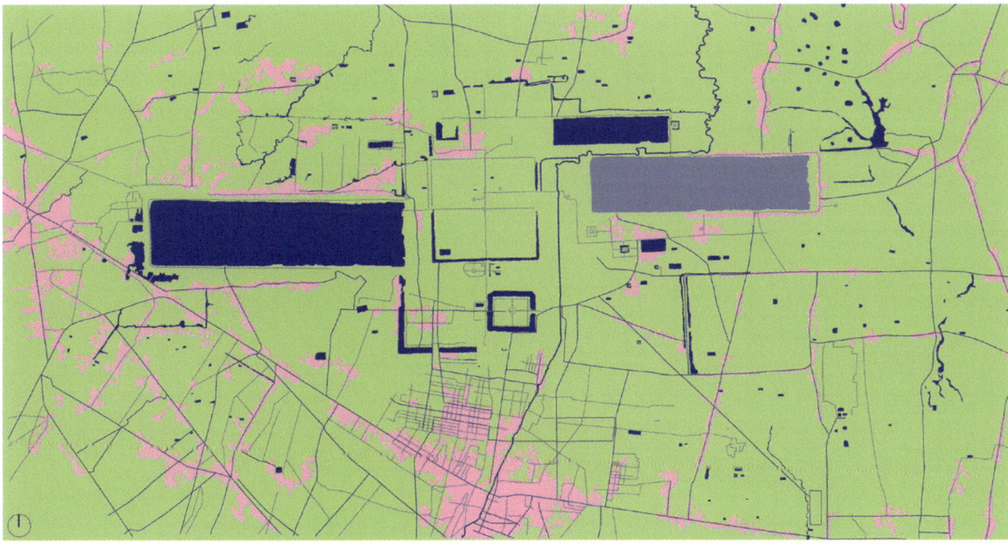

3 - The Khmer Empire was structured by an ingenious water management system that dovetailed with production, sacred and settlement systems, 2022. (Image courtesy of Research, Urbanism and Architecture)

4 - The ancient Singhalese capital, Sigiriya, was designed as a constellation of Buddhist monasteries, village tanks (reservoirs) and major iron producing centers, 2007 redrawn from Bandaranayaje 1993, p.135. (Image courtesy of K. Shannon)

EXTREME INDUSTRIALIZATION AND EXTRACTIVISM, EXCESSIVE ENGINEERING, AND DEVASTATING ECOLOGICAL CONSEQUENCES

In addition to the constant advancement of engineering in irrigation and drainage techniques, there has been a near continuous move to what British geographer Matthew Gandy, has termed the 'hygienist city' or 'bacteriological city', initiated by nineteenth century public health and sanitation concerns (Gandy 2016). The so-called 'Great Stink' of London's Thames River required a more sanitary manner to discard liquid and non-liquid waste. The summer of 1858 saw an excessively pungent stench of untreated human waste and industrial effluent that emptied directly into the city's river and instigated the near universal development of modern 'end-of-pipe' sewage treatment and urban water management systems, not only for waste, but also for stormwater. From that period onwards, water was primarily conveyed in underground pipes: out of sight and out of mind. In line with the exponential urban growth of the last two centuries, ever-increasing quantities of water are extracted from further and further away, distributed through more extensive networks than existed previously, and discharged after use in ever larger scale sewage systems ending in centralized water treatment plants, thought to be the best solution. However, water pollution has become a major threat. In many, if not most, contexts, urban sewage pipes built in the last 150 years and newly built (particularly in Asia and Africa) are both undersized and under maintained. And parallel to the mechanization of wastewater has been the increasing disturbance to natural systems, particularly due to increased sealing of soils in the process of urbanization and its consequential decreased infiltration for groundwater recharge.

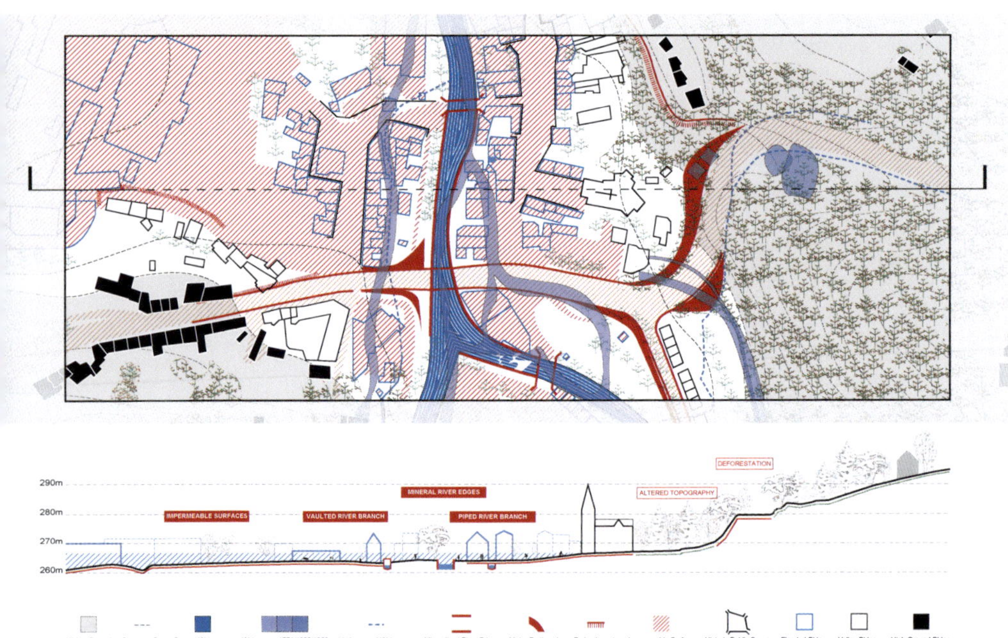

5 - The Vesdre River in Belgium is representative of the industrial-era hard engineering of rivers, where its floodplain was significantly constricted, 2021. (Image courtesy of MaHS/MaULP, KU Leuven)

Over time, industrialization—particularly in Europe—intensified and radically transformed the natural hydrologic cycle and simultaneously decreased water quality. The textile industry was the first to burgeon; factories were developed on riverfronts and relied on water for processing and transport, with the consequence of catastrophic pollution. Over time, ever-larger riverfront and coastal industrial areas became closed off, privatized domains that blocked views to once picturesque valleys and seafronts, making these waterfronts the back sides of public spaces. Factories expanded to include paper, metals, chemical and food industries as well, and intensified the scale and scope of hydraulic change. Railways, a masterpiece of 19th century engineering, meandered along valleys on bridges, impressive viaducts and in tunnels, and on flat coastal plains on raised ground. As a voluminous topographical intervention, railroad and later highway construction led to substantial land filling, to address engineering concerns about water, within natural floodplains. Waterways became more and more artificially constricted. Operating *de facto* as dikes, transportation infrastructure significantly changes basin hydrology—blocking runoff or serving as dams. Once wide and richly braided rivers, and finely articulated deltas and coasts have become ever more narrowed and, in many instances, geometrically engineered and literally hardened.

As industrial and infrastructural development progressed, flood plains continued to dramatically shrink, as lakes and wetlands disappeared due to landfilling of sites for ever-larger industrial platforms and urban expansion. The flipside of modernization, but inherently related to it, is the gradual disappearance of natural wetlands and floodplains (often in colonial, and later in so-called global south, contexts) that were systematically reclaimed, drained, and irrigated depending on the season, to turn them into agricultural areas (Marot 2022). Once an engineered water resource management system was put in place, mechanization and artificial fertilizers allowed for extractive monocultures on an industrial scale—heralding the so-called Green Revolution. The Mekong Delta is an exemplary case of this metamorphosis from a natural delta into a giant mosaic and highly engineered network of paddy fields, commonly labeled as the rice basket of the region. Just as amazing are the vast irrigation networks developed over time in arid and semi-areas, often bringing water from far away watersheds or ever deeper wells, as in California's Central Valley, nowadays the most productive agricultural region in the United States.

In urban areas, substandard and speculative worker housing encroached on liminal spaces between waterways and infrastructure in an endless and almost uninterrupted ribbon development. Globally, complex hydrology became radically simplified and the braided fluvial geomorphology of floodplains all but disappeared as effluents were channeled or buried, islands eliminated, and meanders cut. Once soft watershed lines became further hardened with canal and road construction, and vast impervious surfaces replaced wetlands. From the early 20th century onwards, supply reservoirs for drinking and irrigation water, hydroelectric dams and power plants, further regulated natural water systems, turning entire watersheds into highly engineered territorial systems spanning both countryside and metropolitan regions.

In fact, dams and their accompanying reservoirs became the icon of twentieth century large-scale hydraulic manipulations of landscapes. Increasing in volume and number with each recent generation, they are evidence of a technological capacity that nears hubris (Turpin 2008; Macy 2009). They are considered necessary basic infrastructure for the water and energy supply for ever-larger populations. As such, dams and reservoirs were, and still often are, considered symbols of modernization—from the Tennessee Valley Authority's projects to megalomaniac plans of fascist regime, such as Atlantropa in the Mediterranean Sea. Atlantropa, a project by the German architect Herman Sörgel, envisioned the damming of the whole Mediterranean through gigantic land reclamation made possible by the engineered sea level drop. By taming and regulating wild and irregular natural forces, they turn potential into a supposedly assured promise of development and progress. Tested and perfected in the industrial world, dams and reservoirs were quickly spread worldwide through various imperial enterprises and, in a tangible way, supported nation-building and globalization. The Three Gorges Dam—larger than any before it—is only one of the testimonies of the resolute belief in dams as an instrument of development and the expectation of affluence thought to automatically follow in its wake. Dams and reservoirs illustrate,

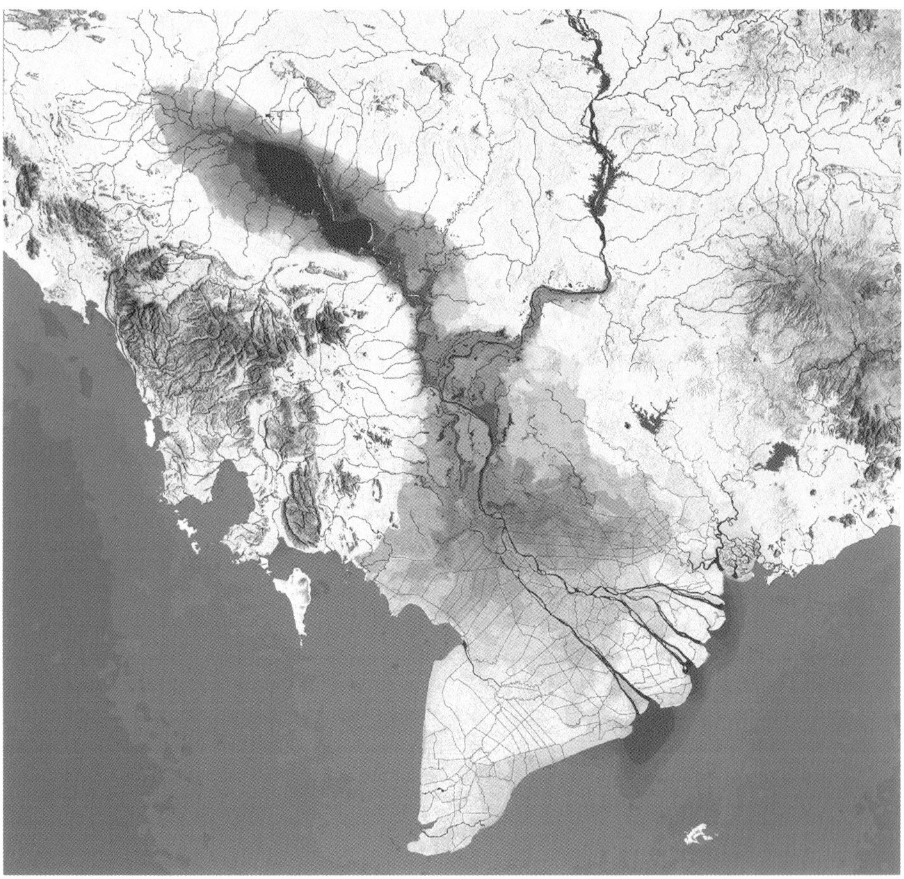

6 - Over centuries, the Mekong Delta, in Cambodia and Vietnam, has been transformed from a vast swamp into a highly engineered and productive territory. Water and flooding regimes remain the register, despite the onslaught of road-based urbanism, 2022. (Image courtesy of Research, Urbanism and Architecture)

as no other construction does, the etymological meaning of infrastructure: the basis of a system of development. However, dams and reservoirs are also highly contested due to their enormous ecological and environmental impacts, as well as the often dramatic social and economic consequences for displaced populations and their painfully unfulfilled expectations of development. The Cahora Bassa Dam on the Zambezi River, the largest dam in Southern Africa, is unfortunately a typical example of this. Built by the Portuguese between 1969 and 1974, this late colonial project delivers electricity to cities, mines and farms in South Africa, while leaving local communities without electricity and with major water supply issues—including major disturbances to the river regime, and hence agricultural activities downstream. The construction expelled, without compensation, local communities from their native lands. The dam was heavily contested before, during and after construction, with major sabotage operations during Mozambique's Civil War (1975-92) that followed independence (Isaacman & Isaacman 2013). In the meantime, as an effect of global warming, water levels in many dams are decreasing at an alarming rate, endangering water supply to whole metropolitan regions such as Los Angeles, and falling brutally short on the promise of assured energy delivery.

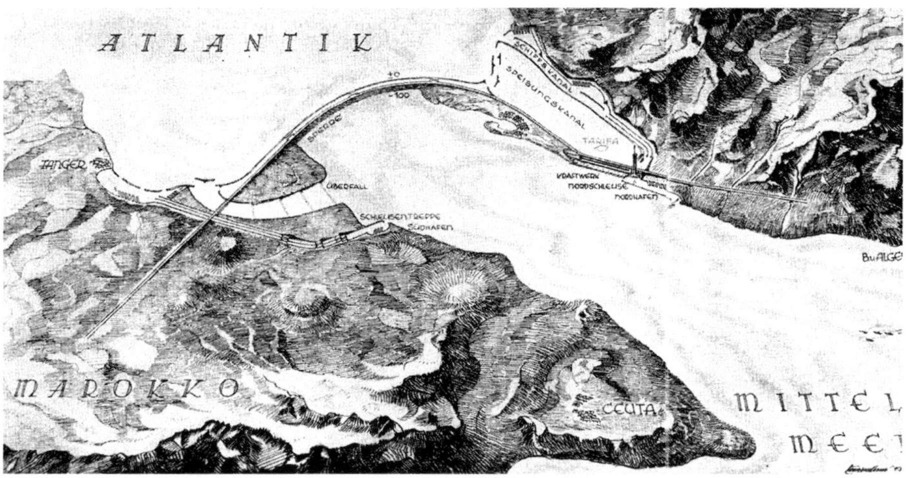

7 - Atlantropa was developed by Herman Sörgel for the Nazi regime over three decades. The proposal was to build a hydroelectric dam across the Strait of Gibraltar—providing not only energy but also vast tracts of land for settlement in land that was once the Mediterranean Sea, 2023. (Image courtesy of Creative Commons)

By the 1950s, when highways became new vectors of development, there was a complementary process of massive suburbanization. The footprints of urbanization have been escalating ever since. It is evident that such sprawling development dramatically disturbs ecologies and has harmful hydrological effects. Geographies were systematically leveled, inconvenient waterways and wetlands covered and water tables dramatically disturbed. Deforestation and the accompanying landslides were accelerated and there has been a sharp increase of the area of sealed land surfaces—all with the cumulative effect of a sharp intensification of stormwater discharge, exacerbated drought and clean water scarcity.

From the local well-known 1970s environmental tragedy of Love Canal (an abandoned canal in the city of Niagara Falls that became a dumping ground for toxic chemical waste) and the 2014 Flint Water Crisis (a public health crisis related to contaminated drinking water for the city in Michigan) to global contamination of the entire water network (with pervasive chemical pollution and microplastics in the sea) to the continued lack of access to safe drinking water for 30% of the world's population (UNICEF/WHO 2016), it is evident that the cascading environmental crises are inextricably linked to growing socio-economic disparities. The era of modernization and industrialization triggered not only an environmental crisis as powerfully and lyrically brought to the public's attention by Rachel Carson's 1962 *Silent Spring*, but also an array of glaring injustices as written by Naomi Klein in *This Changes Everything: Capitalism vs. The Climate* (2014).

THE ANTHROPOCENE, GLOBAL WARMING, AND NEW TENDENCIES

As the world's economy, politics and ecologies become more intertwined, there is not only an increase in inequalities, but also an increase in the frequency and intensity of 'natural' disasters. Water is not accidentally a major agency in all these disasters—floods, landslides, avalanches, massive glacial melts, and more. Asking to what extent the 'revenge' of nature is human induced is answering the question. Over time and through humankind's interventions, the once subtle and ever-shifting gradients of wetness and the trajectories of rivers and wetlands were transformed into categorically distinct areas of dry and wet. The seasonal rhythms of water flows were tamed, and the perils of floods were circumvented, until they no longer occurred: the water stored in glaciers is now melting; water is no longer absorbed in massively increasing deforested areas created by forest fires or by logging; sea levels are rising; aquifers are depleted; and seasonal rain patterns are radically shifting. *It's freezing in LA* might be an amusing magazine title, but the unprecedented floods and accompanying landslides and sinkholes of early 2023 in Los Angeles and California were not amusing but shocking. Clearly, the havoc brought to water systems is part and parcel of the Anthropocene—a proposed geological epoch used to describe the most recent period in Earth's history when human activity started to have a significant impact on global warming and ecosystems.

As the environmental law professor Jedediah Purdy has stated, "humanity has outstripped geology" (Purdy 2015, p. 1). Purdy explicitly refers to water management to reveal how 'wilderness' and 'waste' were (and continue to be) extensively 'reclaimed' for human purposes. In the United States, 'The Swampland Act of 1850' gave public wetlands to those who agreed to drain them, while irrigation laws exchanged ownership for canal building in dry lands. By the later nineteenth century, when it became clear that most private irrigation efforts would fail, the federal government undertook a massive reengineering of the continent's water supplies, moving the precious liquid from mountains to plains, from rivers to fields, until the tellingly named Bureau of Reclamation had hardly left a free-flowing waterway in the West" (Purdy 2015, p. 73-74). At the same time, "nature contained both the wild terrain that called out to be cleared and settled and the beautiful and sublime landscapes that offered symbolic, aesthetic and moral lessons" (Purdy 2015, p. 104). The American Swampland Act isn't so different in aim and consequence than many of the Europe's contemporaneous legislation (for example the Belgian Wasteland Clearing Act, *Loi sur le défrichement des terres incultes*, of 1847) that brought common 'wastelands' such as swamps, heathers, moors, and marshes into development. 'Wastelands' entered the planning discourse since the first European atlas and survey enterprises of the Enlightenment, and soon became a legitimating argument for development in European colonies. Extractive and profitable monocultural domains were developed through the clearing of supposedly unoccupied wastelands, bringing them into 'culture,' by draining and irrigating them.

RENATURALIZATION, LIVING WITH WATER, AND INTERLINKED LANDSCAPE ECOLOGIES

American academics Anuradha Mathur and Dilip Da Cunha situate their design research practice in a critical zone of wetness, between clouds and aquifers, rather than on a surface categorically divided between land and water. According to Mathur and Da Cunha (2014):

> We live in the time of wetness, a wetness that is everywhere before it is somewhere. It aligns us to a design imagination that accommodates fluidity, openness and complexity. It also pushes us to explore the deep-rooted land centric imagination that has disciplined the world around us to subjugate water. It is a world that is unraveling before us. With rising seas, flooded cities, polluted rivers, piling wastes, and widening inequalities, we believe that ubiquitous wetness in place of the land-water binary holds the way forward. It is an exciting pre-disciplinary ground of design by which we re-articulate the past, experience the present, and envision the future.

The contemporary world is at a critical fulcrum—one that is disturbingly divided and environmentally devastated. The myriad issues aligned with water are central to social and environmental justice. The tapping and distribution of water necessitates reconsideration and many of the hard engineering approaches are being questioned and historic methods are being critically reconsidered. The 'settling with water' in the 21st century requires a shifting of scales towards water basin thinking, measuring, designing and acting. With the increase of natural disasters in frequency and intensity, scientists, politicians, civil society, and citizens alike have come to understand that extreme events are no longer exceptional and that there is a necessity to reimagine the nature/culture, water/land and consumptive/productive landscape relationships. Water basins are an interesting and relevant scale to consider with regards to the conceptual and technical understanding of water. This understanding should include what water really is for humankind: both a blessing and a threat, as exemplified by flooding—a phenomena that has deeply rooted negative connotations (mostly for urban citizens) while it is welcomed as means of fertilization by farmers. Water needs to be assessed from multiple perspectives in order to result in new, less invasive responses towards its management.

Currently, unexpected and unprecedented floods come with shocks of devastation. Various government scenario-based flood hazard maps are often woefully inaccurate, with huge mismatches between the flooded areas and the scenario maps, as demonstrated in the water-bomb floods of the Vesdre River in Belgium and the adjacent region in Germany in July 2021, and the global warming induced exceptional floods of the Indus in Pakistan in spring and throughout summer 2022. California experienced its own, completely unexpected version of water bomb rains, causing floods in early 2023, after years of extreme drought and consequential devastating forest fires. All this surely calls for more updated methods of scenario-based hazard maps and risk assessment and the radical rethinking of crisis management. It also suggests the need for stronger alliances between governments and impacted citizens, with new forms of coordination. Risk assessment and crisis management are only the first step of the needed fundamental rethinking and subsequent 'settling' with water. The scale of watershed management needs to be thoughtfully re-defined and to serve as a base for the restructuring of territorial occupation. A generative co-presence and balance of agricultural, urban/rural and natural systems such as forests has to replace the pollution, disorder and mono-functionalism of the industrial era. The intended transformation must address water impacts such as long droughts and devastating floods. It is clear that such a 'settling with water' can anchor water resource management on the renewed water cycle, and that human occupation can be rethought in light of the new nature of water, with the increased frequency and intensity of extreme situations. A non-nostalgic reflection and deep understanding of earlier forms of 'settling with water' could provide valuable lessons for the future. Water is already reconquering the design agenda and there are three broad design strategies outlined below that require a crossing of disciplinary and scalar boundaries, out-of-the-box thinking, and new water condition design thinking.

Strategy 1:
Renaturalization, Reverse Engineering, and Managed Retreat

The explicit recovery and choreographing of ecological processes are considered paramount. At both the territorial and local scales, there is an opportunity to renaturalize water systems, particularly river courses. For decades, their straight jacketing in concrete has artificially controlled flows. There is a growing tendency to restore more 'wild' flows and (re)create spatial qualities of degrees of wetness. The release of rivers and their tributaries from their hard-engineering confines would be a first step to allow water flows to once again sculpt the land and reinstate flood plains. The rewatering of the landscape would entail not only attention to ecological issues, but also demand a cultural shift. River valleys, slopes and plateaus can be reconsidered as one system that requires requalification through restoring ecologies across sections and scales. The waterways would need to be perceived as a series of wetland gardens, for occasional use and enjoyment of humans, but also and primarily, as a home for a wide range of non-human plant and animal species. As various areas of the world de-industrialize, there is an opportunity to change priorities—away from industrial productivity and its attendant pollution and human exploitation, towards regeneration. Ecosystems are in dire need of being healed, which requires not only a paradigm

8 - Vietnam's ancient Imperial city, Hue, is embedded in a water system that spans from its high mountains bordering Laos to the East Sea, 2019. (Images courtesy of Research, Urbanism and Architecture)

shift, but also simply time: time to reestablish native plant communities, time to restore soil and water health, time for forests on steep slopes to grow, mature and expand, time for hedges and other boscage landscape elements to weave ecological webs and water storage systems, time to develop new expertise, perhaps time to simply adopt new healthy ways of life supported by the economies generated by a careful use of the resources, time to learn to adapt, and time to develop new urban morphologies, new landscape fabrics and innovative architectural typologies. Fragility, temporality, unpredictability, and beauty would be the new assets—far from a past of harnessing waters for economic return or simply as the backsides and dumping grounds they have become during industrialization. In some postindustrial instances, the rewatering of river braids would be able to naturally 'solve' part of the deurbanization 'plague'. Undoing existing dams is fortunately already occurring and should gain momentum in the near future. This is another radical approach that is gaining ground in river restoration projects, as evident in Washington State's Elwha Dam removal of 2011-2012, as changing water regimes due to global warming has rendered many dams incapable of delivering envisioned energy and/or potable water.

Strategy 2:
Living With Water

For all too long, humankind controlled water and almost attempted to rid themselves of it: put it behind dikes and dams, pipe it, or bury it. It was out of sight, out of mind. Gradients of wetness were categorically turned into dry or wet landscapes. And slowly, age-old typologies such as stilt and floating buildings were systematically replaced by *terra firma* construction. Water disappeared out of sight in cities. The increase in the frequency and severity of flooding, compounded by sea level rise, is leading to the reexamination of old typologies in addition to more high-tech versions.

9 - Washington State's Elwha Dam was built in 1913. Its deconstruction began in 2011 (L) and was complete by 2012 (R). (Images courtesy of Photographer Ben Cody, Creative Commons)

Reversing water from *de facto* backsides of cities into spacious and open, small or large, natural centers of the settlement, is a way to induce potentially new and vast public spaces. These were never fully developed during the Industrial and Urban Revolution, and will make human settlements constructively re-engage with their ecological environment and its basic resource—water. There is surely an urgent need to give water again an evident presence in the everyday environment instead of the technical out of sight/out of mind approach that prevailed during most of the 19th and 20th centuries. Such ecological re-engineering can be complemented by the renewal of public spaces which re-engage with water such as rivers, lakes, wetlands and coasts, and simultaneously afford places of collective memory. In post-industrial areas, ruins can become part and parcel of the new and dynamic urban water landscapes. New building typologies with, for example, elevated ground floors or floating capabilities, can allow for seasonal flooding, and for differentiated ways of building on dry, marshy or wet ecotones. Living with water requires a fundamental shift from a linear 'extract, consume and dispose (and if really necessary clean)' water approach, to a more circular approach where reuse and regeneration take more prominent roles, working within the logic of water cycles in watersheds rather than the denial of these natural systems. This shift requires rethinking across scales, keeping in mind Mathur and da Cunha's 'wetness is everywhere before it is somewhere' (Mathur and Da Cunha 2014). This provides a multitude of opportunities and possibilities to conceive new nature-based solutions, even beyond the water sensitive urban design tools (WSUD) that are already circulating in today's professional practices worldwide. There is an enormous need for the (re)creation of water bodies to break through the quick linear discharge of stormwater and wastewater through underground pipes.

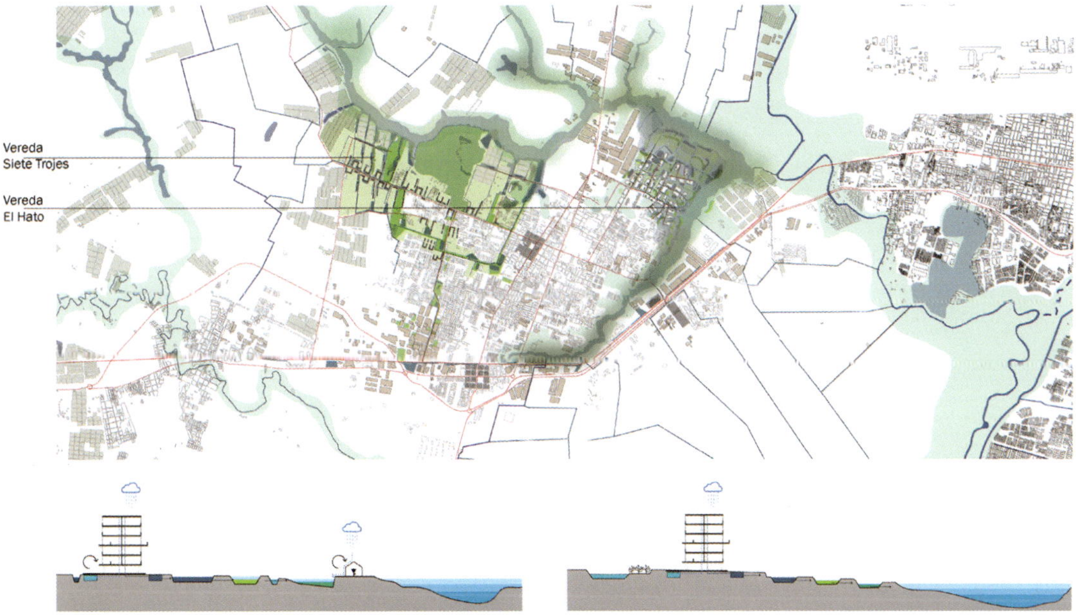

10 - Design research in the Sabana de Bogotá re-articulated the nature and city relationship through the water structure. New housing typologies were choreographed within a preserved wetland, 2017. (Image courtesy of Claudia Lucia Rojas Bernal)

Strategy 3:
Interlinked Landscape Ecologies

Water bodies must be understood, transformed, and regenerated as a series of interdependent ecologies, across lagoon, coast and valley sections, across the segments of lower, middle, and higher basins, and across scales. Systematic exploration of water through a sectional analysis, often with upland systems of forests, followed by agriculture on plateaus and down to valleys with floodplains, would allow for the conception of interventions to rebalance occupations that align with a natural regeneration of water cycles. The rebalancing of agriculture and forestry usually allows for the systematic reweaving of landscape elements and includes nature-based solutions for water retention and storage. Besides restoring the sponge-capacity of plateaus and slopes, systemic afforestation across entire valley sections would regenerate

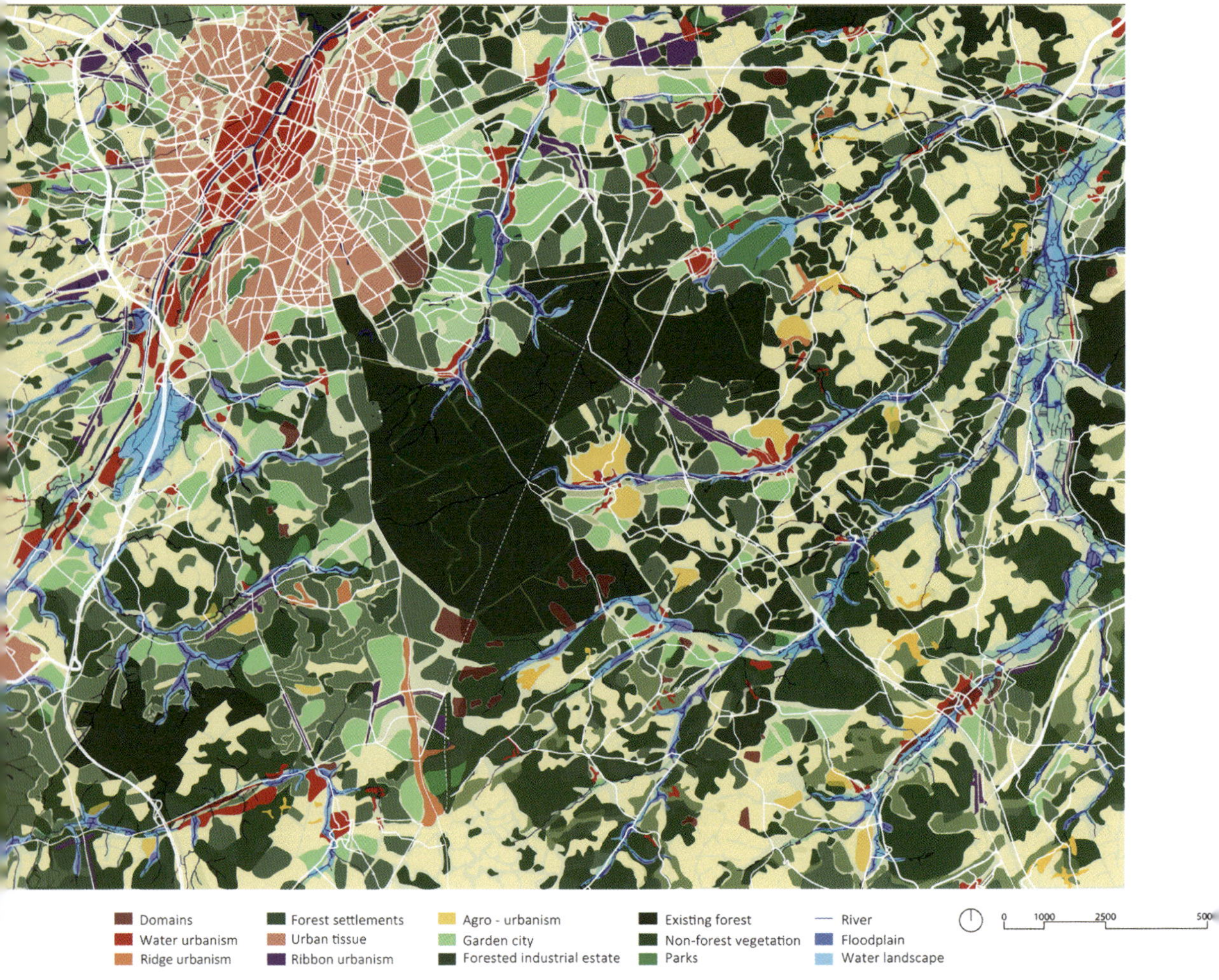

11 - The Brussels' Sonian Forest is envisioned as a national park that strengthens the ecology of the forest and its intertwining with water structures, 2018. (Image courtesy of Research, Urbanism and Architecture)

broader ecologies and create a healthier living environment, again for humans and non-humans alike. Regularly spaced, experimental and counterintuitive floodable forest patches within braided floodplains can protect downstream settlements and achieve a softer modulation of water flows along the entire sectional area during storm events, while being part and parcel of the long-term natural park system that would terminate in the floodplain.

NECESSITY OF DESIGN

Humankind is living in a perilous moment, with crises and disasters following quickly one after the other. It could easily give way to pessimism, but in taking clues from what writers Rebecca Solnit (2010) and Arundhati Roy (2020) have articulated, in the wake of disasters there are moments of altruism, resourcefulness, and generosity amid the grief and disruption. These moments are portals to a new vision of what society could become—one that is more collaborative, more adaptive and more locally-embedded. This augurs new water relationships.

The increased frequency and intensity of extreme weather events across the globe make it evident that there is a need for a radical re-imagining of a host of relationships—nature and culture, ways to live together with one another as well as with other-than-human species, global and local, built and unbuilt territories. Humankind needs new ways to settle, in terms of literally constructing settlements as well as 'settling' with the environment, with nature in general, and with water in particular. The necessity of design and the role of architecture, landscape architecture, and urbanism are central to the envisioning of possible futures, particularly regarding aspects of social and environmental justice. The centrality of the fields needs to be urgently reclaimed.

Changing Water Cycles in the Alps

Silvia Picone

When the well's dry, we know the worth of water.

—*Benjamin Franklin*, 'Poor Richard's Almanack, 1746'

Water is essential for human life and settlement. What we are witnessing in terms of the degradation of water supply and water quality as a result of pollution and climate change requires a new focus. In order to address these new risks, action by all is fundamental. Starting from a single drop of water and connecting it to the currents and structures it forms, just like extrapolating from the individual to world politics, we must re-adapt the way we manage this element, moving towards a sustainable use. This challenge is very visible in Northern Italy, where the recent impacts of climate change in the Alps have led to changes in water supply, resulting in a water crisis.

Water is a human right.[1] Water is: 'life', which we use to drink and clean, and to nourish ourselves; 'food', which exists throughout the food chain thanks to water; 'industrial processes', including the extraction of raw materials, which require water to function; 'economy' and all goods, such as fabrics made from the cotton plant, which depend on water to be created. Local responses to water supply and water quality must include public debate and transparent science-based legislative guidelines.

NORTHERN ITALY: WATER, CLIMATE CHANGE AND LOCALIZED IMPACTS

The *Intergovernmental Panel on Climate Change* considers the Mediterranean region a 'hot spot' of climate change, warming 20% faster than the global average (Hilmi et al. 2022). From 1880, Alpine temperatures have risen at a rate about twice the northern-hemispheric average, with an overall increase of 2°C in the mean annual air temperature (Auer et al. 2007), and a particularly strong warming over the past two decades. This affects Italy by an increase in the magnitude and frequency of extreme events. The impacts include melting glaciers and rising sea temperature, causing changes to the life cycle of water.

Italy, as is much of the rest of the world, is also profoundly affected by events of combined hydrological, geological, and hydraulic volatility. The impacts expected from climate change and extreme weather will contribute to intensified instances of land instability, amplifying and aggravating an already critical situation and increasing the frequency and the magnitude of natural hazards that include rock

[1] The human right to water was recognized as a fundamental right by the UN in 2010.
[2] Almost 94% of Italian municipalities are affected by areas with high or and very high landslide hazard risk.

1 - Geohydrology—water among the rocks at the Enna river in the Taleggio Valley. Sofia Malone. *River Rocks*, 2017. (Image courtesy of artist)

falls, landslides,[2] avalanches, debris flows, and floods. These geo-hydrological threats will increasingly cause physical damage, impact the safety of the population, and impact biodiversity.

Not only do water system events driven by climate change have impacts on life, but they are also very expensive to society. Between 1951 and 2009, the total cost of damage caused by landslides and floods in Italy amounts to over 52 billion euros, while the financial requirement to mitigate the risk of landslides and floods is around 40 billion euros (Trezzini et al. 2013). By comparison, the Italian state invested roughly the same in public education in 2017 (66 billion euros) and almost double in the health sector in 2021 (135.5 billion euros). In more recent years the costs incurred by the Italian state coffers and local communities from landslides, floods and avalanches has increased from approximately one billion euros per year to 1.5 billion (Trigila et al. 2015) resulting in greater state spending in response to the emergency, which could be less if we migrated toward investing in long-term prevention and mitigation.

In order to understand the effects of climate change on water systems, this essay focuses on how climate change directly affects the water cycle in this Prealpine northern Italian region, foregrounding an example of the magnitude of global changes in a detailed local situation.

RISKS FROM THE CRYOSPHERE: GLACIAL RETREAT AND PERMAFROST DEGRADATION

The Taleggio Valley is situated in the Prealpine region of Italy. The cryosphere is retreating through mass losses of ice and snow due to its high sensitivity to climate change.[3] In turn, these changes have a direct influence on the climate, allowing greater levels of solar energy to be absorbed by land or water, which in turn leads to further warming.

Currently the cryosphere provides reliable water supplies for tens of millions of people, but the future reliability of this resource is at risk. The average temperature of the Alps has risen by 2°C since the end of the nineteenth century, which is roughly double the global average increase.

2 - Images showing the retreat in ice extent on the Forni Glacier, Valfurva (So). Photo comparison shows the glacier in 1949 (L) and 2020 (R). (Images courtesy of Comitato Glaciologico Italiano, photos by A. Desio and S. Perona)

Mountain glaciers are recognized as excellent natural features for investigating climate variability across decades and centuries. They visibly respond to variations in temperature and precipitation, and therefore are recognized as one of the best terrestrial indicators of climate change.

The glaciers of the Alps tell the change story. Since the end of the Little Ice Age (circa 1850 to 1980), they underwent a general retreat, losing half of their volume. However since the 1980s they have been disappearing faster,[4] with a marked accel-

[3] The cryosphere is the part of the Earth's surface characterized by the presence of frozen water, and includes components such as snow, glaciers and permafrost which are extremely sensitive to recent climate change as they are subjected to melting and thawing conditions.

[4] A further 10-20% of their volume has disappeared since the 1980's (Diolaiuti & Smiraglia, 2010)

eration in retreat observed since 2003 (D'Agata et al. 2020). The most reported changes are: the glacial front retreat of tens of meters, an increase in debris cover, the development of large hollows on the glacier surface, glacier fragmentation, and enhanced glacial flow.

The highest reduction in surface area is seen in the small glaciers which represent roughly 80% of the glacial fields in the Alps, and most importantly make a vital contribution to water resources.[5] Models show these small glaciers will disappear completely in the next few decades, while the larger ones are expected to experience a further reduction in volume of between 30% and 70% by 2050 (Smiraglia et al. 2015).

The loss of the large and small glaciers will have a profound impact on the region. Local agriculture will lose a major source of irrigation, while the five hundred-plus hydroelectric power plants which depend on the melting of alpine glaciers to provide water, will lose power generation capability.

Climate change also reduces periods of snow and increases the frequency and intensity of extreme precipitation events, leading to the progressive degradation of permafrost. Permafrost is defined as ground that remains at or below 0°C for two consecutive years, and is characterized by typical landforms such as rock glaciers, ice-wedge polygons, and frost mounds that are closely tied to ground ice. Rock glaciers can store water in solid form (ice) as well as in liquid form (groundwater within the unfrozen base layer), and they act as shallow groundwater reservoirs for the Alps (Guglielmin et al. 2018).[6]

Under current trends, it is expected that glacier ice will disappear in the Italian Alps by the end of this century, while the fate of permafrost ice within rock glaciers is still uncertain and poorly understood.[7] It is estimated that the ice present in rock glaciers is more resilient to global warming than glacial ice thanks to the characteristic coarse debris that covers the rock glaciers and acts as an insulator. The importance of potentially more resilient permafrost as an alternative water source to glaciers cannot be underestimated.

Changes in precipitation patterns affect the cryosphere and initiate several concerning positive feedback cycles.[8] We are now witnessing short and intense rainfalls with contrasting periods of drought. These result in the soil becoming impermeable due to saturation, increasing the hydrogeological risks that can create landslides and similar events, as well as presenting a risk to the quality and availability of water resources

[5] Surface areas <1 km^2

[6] Rock glaciers are often located next to tongue-shaped bodies of frozen debris with interstitial ice cement, ice lenses, or a core of massive ice that moves slowly down a slope under the influence of gravity.

[7] Loss rates of glacier ice in the Alps are on the order of 800-1000 kg m^2 per year, while rates for subsurface permafrost ice loss are 1-2 orders of magnitude lower.

[8] The process known as positive feedback in relation to the climate is when climate change causes the effects of an initial alteration to be amplified. For example, in this case positive feedback would increase rather than reduce the initial rise in temperature.

in northern Italy, exacerbating the strong competition for water resources. Locally this affects the amount of water in lowland areas below the prealps, and impacts the economic sectors in those areas that rely on water, such as agriculture, hydropower production, industrial production, tourism, and river navigation.

THREATS TO LOWLAND FRESHWATER SYSTEMS

Italy has a particularly large reliance on freshwater. According to the *Italian National Institute of Statistics* (ISTAT), in 2015 Italy was the highest user in the European community of freshwater in the residential arena.[9] Many other sectors also rely upon freshwater supplies, in particular agriculture, the industrial sector, and manufacturing.[10]

The wildlife dependent on freshwater is also under threat. In the past thirty years, freshwater biodiversity in Italy has declined faster than marine or terrestrial biodiversity (Jenkins 2003). Species living in freshwater ecosystems in the Mediterranean are highly specific.[11] These ecosystems are particularly vulnerable to climate change,

3 - Matthew Wong. *Freshwater infrastructure for hydropower in the Orridi of the Taleggio Valley,* 2018. (Image courtesy of artist)

[9] The total fresh water consumption in Italy in 2015 was approximately 21.6 billion m^3, with an amount of drinking water supplied to final users in households of 4.78 billion m^3, accounting for 22% of total water consumption.
[10] Between 2012 and 2015, the overall water withdrawn in Italy for agricultural uses represented about 50% of the total freshwater withdrawals. The civil sector (which includes residential water) withdrew about 36%, and the industrial sector and manufacturing about 14%, with strong differences between the different regions of the country.
[11] In Italy, the percentage of endemic species (those restricted to a specific geographical area) in the freshwater ecosystems exceeds 10% of the total endemic species in Italy (Stoch 2003–2004).

as it alters stream flows and water quality, with dramatic impacts on freshwater biota. In addition, freshwater ecosystems are expected to experience alterations, like phenology changes (alteration on timings of cyclical events), upward shifts (elevational range shifts of mountain species), and invasion of alien species (non-native to an ecosystem), with a consequent increase of competition and predation leading to reduced species richness. This could represent a special vulnerability in alpine freshwater ecosystems.

With increasing temperatures and decreasing precipitation in arid and semi-arid parts of the nation, the biggest danger is the expected decrease in water inputs. Adding to this is the increase in evapotranspiration, which leads to saltwater intrusion into coastal fresh-water beds, and the loss of wetlands. These result in severe imbalances in the wetlands of the coastal zone with changes in salinity and hence in the related biotic communities (Fenoglio et al. 2010).

THREATS TO RIVER SYSTEMS IN ITALY

The winter of 2021-22 had the lowest recorded snowfall in the last twenty years and was one of the ten driest winters in the last hundred years. While changes to the cryosphere have an impact on freshwater in Northern Italy, the decrease in snowfall in the Alps also has direct effects on the river systems of the plains. Rivers play a key role in water circulation and geochemistry. Climate change-induced changes to the freshwater supply from the cryosphere, can cause irregularities in large river systems which results in critical consequences.

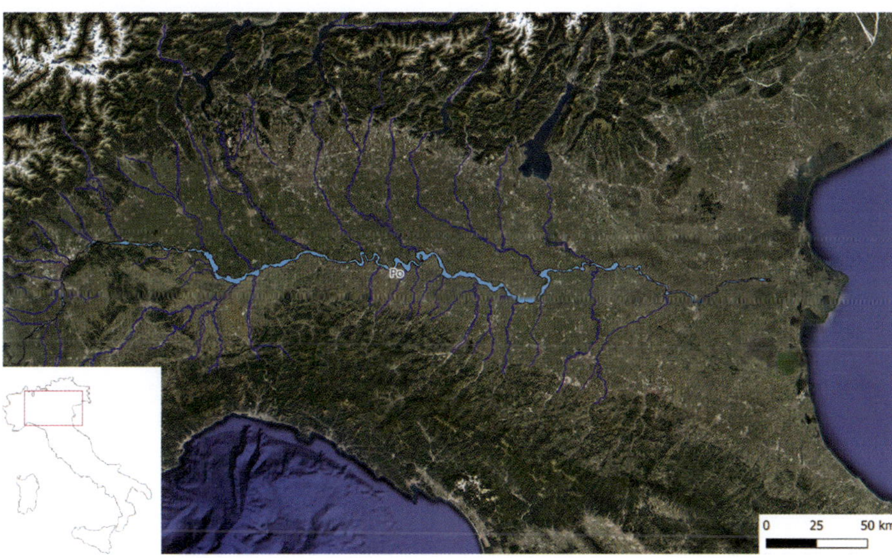

4 - The Po river with its tributaries. The Po River basin is one of the seven Italian river basin districts established by decree of the *Ministry for the Environment*, Land and Sea. (Image courtesy of Google Earth, 2022)

The longest Italian river, the Po, nourishes the fertile plains of northern Italy, which are responsible for some 40% of Italy's GDP and produce around 30% of the national agricultural production, including wheat and rice.[12] In the summer of 2022, Italy declared a state of emergency in five Northern regions due to the worst drought in the last 70 years.[13] The water level was so low that remains from World War II were discovered, and a wall of a medieval village emerged. Towards the river's mouth, the Adriatic Sea rose about 19 km inland from the Po estuary, burning crops, salting drinking water, and causing a serious problem for biodiversity. There was a large reduction in freshwater availability and an increase in flood risk due to the expanding sea. It also caused changes to the surface water salinity affecting the chemistry of the Mediterranean Sea and impacting coastal marine ecosystems.

Continued water shortages in the future, combined with increasing water demand for agriculture and tourism, will have severe effects on local ecosystem services, with consequences such as compromising soil fertility and increasing the risk of forest fires. As seen with the Po River, one of the most significant consequences of climate change is the alteration in stream flow regimes by changing the quantity and timing of river flows, which are essential patterns for the existing biota of river systems. In combination with prolonged drought, this leads to a progressive shift from permanent to temporary water bodies.

5 - Casalmaggiore Port on River Po, 2022. (Image courtesy S. Ravelli)

[12] The Po River flows 650 km (407 miles) from the Alps in the northwest to the Po Delta in the east, rushing out into the Adriatic Sea.
[13] For more than three months in the summer of 2022 there was no rainfall, the average river level lowered by seven meters.

THREATS TO LAKE WATERS IN ITALY

Another key feature of Northern Italy is its large freshwater lakes. These are hugely important water reserves for humans, and support valuable socio-economic systems. They support tourism, recreational-environmental activities, transportation, ecosystem services, fisheries, hydroelectric production, irrigation, and drinking water supply. However, the current equilibrium of all these benefits is highly threatened by human-induced water level fluctuations. Recent climate change impacts compound the problem, and along with pollution levels, the impacts of existing dams, and other anthropogenic impacts, have led to the dysfunctionality of lake ecosystems and alterations in the composition of the biodiversity.

Lake Maggiore is the second largest lake in Italy, half-shared with Switzerland.[14] The level of the lake has been regulated through the Miorina Dam since 1942 to meet the needs of agriculture and industry, manage drought events, and produce hydroelectric power for Italy and Switzerland. The quality of Lake Maggiore water has been extensively studied, with a well-documented history of chemical-organic pollution and eutrophication (excess of nitrogen, phosphorus and other phyto-stimulating substances) from domestic and industrial activities (Manca et al. 2007; Salmaso et al. 2007). Thanks to these studies and concerted actions, the water quality of Lake Maggiore has improved significantly over the last few decades.

6 - Lake Maggiore, Italy. A view of the highly developed shoreline, looking north-west from the south-east, 2019. (Image courtesy of A. Bottelli)

[14] Lake Maggiore has a surface area of 212 km^2 (144,8 mi^2), a maximum depth of 370 m, and volume of 37.5 km^3 (89.9 mi^3).

However, the biological and physical structure of some smaller lakes is being affected by global warming, resulting in cyanobacterial blooms.[15] Cyanobacteria are organisms found in most aquatic and terrestrial environments. Under certain conditions, the cyanobacteria present in the water can multiply quickly enough to color the water and form surface foam. The high presence of cyanobacteria is in many cases linked to the presence of excessive quantities of nutrients such as phosphorus and nitrogen (eutrophication[16]) and is attributable to human activities such as civil, industrial, livestock and agricultural discharges. Their blooms are increasing, in terms of coverage and frequency, due to eutrophication and long-term climate change. An excessive presence of cyanobacteria can have negative consequences for the use of the water, both for recreational and bathing purposes and as drinking water, since the bacteria are capable of producing toxins with various negative health effects.

EUTROPHICATION PROCESS

Due to climate change, the temperature of many lakes has increased significantly in recent decades, reducing the circulation of water in the lake's water column (Adrian et al. 2009). As a result, lower oxygen concentrations in the deeper parts of the lake lead to anoxia which affects the nutrient replenishment of the higher waters and impacts the function of the lake's entire ecosystem (Anneville et al. 2005; Salmaso et al. 2007). Anoxic waters are areas of water devoid of dissolved oxygen. Even short-lived anoxic events can cause mass mortality of fish and benthic organisms.[17] Collectively, anoxic and hypoxic events can cause large reductions in fish abundance, diversity and harvesting in affected waters and can contribute to an overall loss of biodiversity.[18]

7 - Eutrophication process in a pond due to an overabundance of nutrients. Christian Fischer. Bloom of cyanobacteria, 2017. (Image courtesy of Creative Commons)

[15] In the period from 1965-2010, the lake waters were recorded as having an increase in temperature in the 0-30 m layer of roughly 1.4 °C resulting in an increased occurrence of cyanobacterial blooms (Callieri and Piscia, 2002).
[16] Eutrophication is the increase in the rate of concentration of phosphorus, nitrogen, and other nutrients to an ecosystem.
[17] Benthic: from benthos, which is the community of organisms that live on the sediments of aquatic habitats.
[18] When dissolved oxygen concentrations are close to zero it is called an anoxic condition, while if oxygen is <2.0 mg L-1 it is an hypoxic condition.

The decreased snow cover and changes in precipitation, changing water levels, altered hydrologic cycles and disruptions to hydrological connectivity and habitats directly impact aquatic biota. However climate change and drought can also lead to future indirect impacts on biodiversity from the threat of landscape transformations, increases in river morphology alterations, increases in the construction of dams for water supply stability with the consequent river fragmentation it creates, worsening water quality, and changes in food resources.

FUTURE SOLUTIONS: LEGISLATION AND DIRECTIVES AND ITS IMPACTS

In 1976, Italy passed the Merli Law, one of the first important steps in the attempt to protect water from pollution by regulating polluting substances that are discharged into waterways and setting the allowable concentration levels.

Later, with the Galli Law (Italian Law No. 36/1994), a substantial process of modernization and a reorganization of the water sector was introduced. This legislative action focused on the provision of water resources and established

8 - Natural drinking water springs and public fountains in Taleggio Valley, 2018. (Images courtesy of J. Anholt)

that all surface and underground waters are public, and that human consumption prevails over other forms of water consumption. A comprehensive legislative regulation of water arrived in 1999, with the Legislative Decree No. 152, which addressed the safeguarding of water from pollutants, and addressed the treatment of wastewater to protect from nitrates deposited from agricultural activities. This legislation represented an all-encompassing framework for addressing pollution, recognizing the need to address the recovery of polluted water bodies in order to see an improvement in the overall state of water resources, and to pursue a sustainable long-term path towards guaranteeing healthy water resources. This legislation introduced systemic changes at the administrative level, marking a significant departure from the local management principle. Importantly this overcame the managerial fragmentation that had characterized the administration of the water sector until then. Despite the extremely slow operational implementation of the law, it was a provision of great importance, since it initiated the restructuring of the Water Services in Italy.

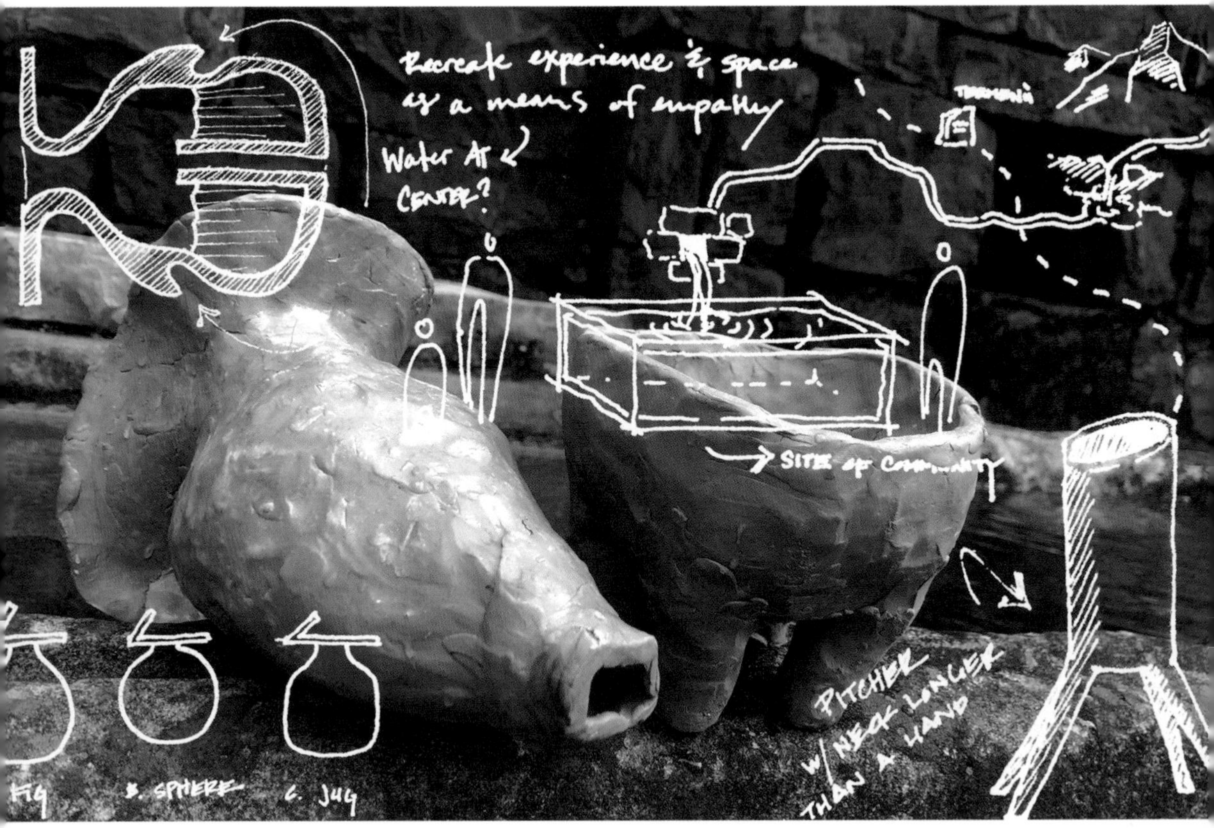

9 - Diagram of public water infrastructures and the vessels made in response. Sam Maddox. *Sip, Serve, Share: Ceramics of Symbiosis*, 2018. (Image courtesy of artist)

In addition to Italian governmental action, protecting water resources is one of the priorities of the European Commission, and their stated goal is to see polluted waters cleaned, while clean waters are kept clean. The *Water Framework Directive* 2000 (WFD) is the most important, far-reaching, water legislation at the European level, and it was implemented in Italy in 2006. It established eight hydrographic districts in Italy, and required each district to prepare a management plan. This was an innovative approach to implementing the European legislation on water, from an environmental and administrative-managerial point of view. The directive set ambitious objectives such as preventing qualitative and quantitative deterioration, improving the status of water, and ensuring sustainable use based on the long-term protection of available water resources. At present, Italy has yet to proceed with the identification of the types of watercourses that would allow the Water Framework Directive to be implemented, as there is no agreement regarding the methodology for this identification. This will lead to a timing delay which will have repercussions on the implementation of the monitoring and management plans. Full implementation of the Directive appears to be still a long way off. In Italy there are many steps to complete and above all it is necessary to correct much of what currently exists.

Looking further afield to the global scale, the 2030 Agenda of the United Nations 2015 Sustainable Development Goals (SDG) highlights the importance of water resources for sustainable development. In particular SDG 6 is dedicated to water and discusses the vital need for improved drinking water. It also provides the framework for improved sanitation and hygiene, and notes that clean water is key to improving health, expanding education, and reducing poverty. It articulates additional targets such as protecting and restoring water-related ecosystems, improving water quality, and reducing water pollution, especially the type caused by hazardous chemicals. Sustainable water management and access to safe water and sanitation were clearly recognized as essential for increasing economic growth and productivity, and for leveraging existing investments in health and education. These objectives have a central role in the sustainable development program. Freshwater resources are in fact essential for health, food security and energy production; if well managed, they can contribute to the fight against poverty.

All the legislative activities noted above were seminal in recognizing that the health of the natural environment contributes to the management and regulation of water availability and water quality. These actions represented an understanding that strengthening the resilience of watersheds also protects against environmental disasters. Water scarcity undermines food security, while improving water management makes national economies, agriculture, and food sectors more resilient in the face of climate change. Actions at all levels, from individual to national, are needed to achieve a sustainable future in light of the local water challenges, such as those of the Alps and the challenges faced by many countries today.

The individual actions that everyone can do include the following:

› Drink tap water, if potable.
› Save water by taking short showers and not bathing. A bathtub holds 100-160 liters of water, while a 5-minute shower consumes 75-90 liters.

- Quickly intervene in the event of water leaks in the taps, in the toilet drain or in the pipes. Just think that a tap that leaks about 30 drops per minute wastes 200 liters of water in a month (24,000 liters in a year).
- Stop polluting: do not throw food waste, waste oil, medicines and chemicals down the drains.
- Buy local foods and those made with less water.
- Collect rainwater and use it, for example, to water the plants.
- Only use washing machines and dishwashers with full loads.

Another problem exacerbating the problem of water shortages in Italy is water transport. In fact, 42% of the water transported is lost to leakage in the piping system.[19] In comparison, in Germany the losses are at 6.5%, in England at 15.5% and in France at 20.9%. Addressing the water lost during transport could significantly help with securing adequate water supplies. To counter this infrastructural criticality of water leaks, funding has been provided for new technologies and resources capable of solving the problem in the PNRR (Piano Nazionale di Ripresa e Resilienza). In total, 900 million euros have been allocated for 33 interventions aimed at reducing losses of drinking water in the aqueduct network. By 31 December 2024 approximately 45,500 kilometers of pipelines for potable use will be equipped with innovative instrumentation and control systems for locating and reducing leaks, promoting optimal management of the water resource, reducing waste and limiting inefficiencies, while improving the quality of the service provided to citizens. The project envisages that in March 2026 these interventions will be extended to approximately 72,000 kilometers of pipelines.

THE WATER FOOTPRINT: A PUBLIC ACCOUNTING TOOL

The water footprint is a tool which can be used to understand how we consume water both indirectly and directly: it can be measured for a single process, a product, or for an entire company. It is a measurement of the amount of freshwater in the volume of water consumed and/or polluted to produce each of the goods and services we use. The calculation includes the direct water use of the process, product, company, or sector used by the individual, plus the indirect water footprint. It is the summation of the water footprints of all the products consumed. It also encompasses water consumption and pollution throughout the full production cycle from the supply chain to the end-user. It is composed of three components: green—rainwater consumption, blue—surface and groundwater consumption, and gray—virtual water for dilution of the polluting load of a process. These categories provide a comprehensive picture of water use by delineating the source of the water consumed and the volume of freshwater required for the assimilation of pollutants.

[19] According to the data published by ISTAT for Italy, of the total of 8.2 billion cubic meters of water introduced into the system, the loss is around 3.4 billion, with the highest value noted in the South and the islands (45%), and the lowest in the North (29%).

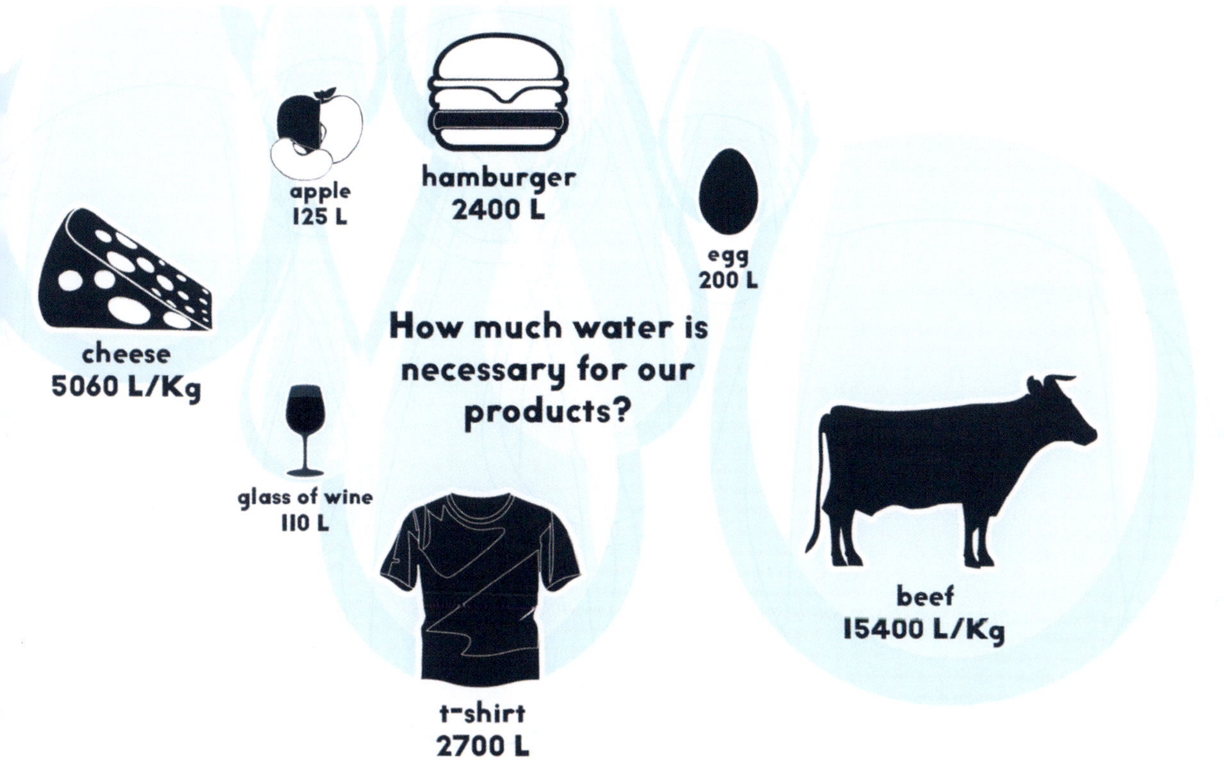

10 - Water footprint (in liter or liter/kilogram) of crop and animal products. (Data from Water Footprint Network, WFN, accessed 2023 - image redrawn by NAHR)

A product's water footprint is the summation of the water footprint of each step, or process, required to produce that product and it is an index of the pressure that the product has put on freshwater resources. Every product we use, from our clothes to the food we consume, needs water for its manufacture. For instance, the water footprint of 200 grams of beef corresponds to 47 eight-minute showers and according to the United Nations Organisation for Food and Agriculture (FAO), to produce a liter of milk, 1,000 liters of water is needed (Hoekstra 2011). Quantifying and monitoring the water footprint of food consumption patterns is important to support the formulation of policies towards sustainable diets. Using this tool it is possible to create more resilient food systems and sustainable diets.

CONCLUSION

Reflecting on the water cycle shows us how rights, laws, environment, and economy are intrinsically interconnected. For this reason March 22nd has been declared *World Water Day* with the aim of raising awareness among world institutions and public opinion of the importance of reducing water waste and adopting behaviors aimed at combating climate change. The interconnected nature of water means that changes on the global scale can have powerful and potentially devastating impacts to cultures, nations and communities on the local scale. The impact of climate change on the cryosphere in the Alps is a prime example of how a global trend puts the water supply of local communities at risk.

In Northern Italy, hydrological changes require a change in the way the environment and land are used. One way to safeguard the water cycle is by re-naturalizing cities and territories. In the presence of trees and forests, water is retained longer from dropping from the foliage to the ground. This does not occur where there is no vegetation such as in heavily cemented cities and in agricultural areas with unsustainable cultivation techniques. In fact, cementification of soils and construction results in the reduction of the soil's ability to absorb rainwater, increasing the risk of phenomena such as flooding and inundations.

11 - This installation responds to the cultural and physical memory of water, and how we are linked to it through our actions, bodies and thoughts. Luisa Brando. *Water Atlas*, 2018. (Image courtesy of artist)

It is necessary to pursue agriculture that allows the soils to retain water by restoring their natural porosity, favoring vegetation that retains water and humidity and that through the roots, helps to infiltrate the water into the subsoil, increasing the fertility of the land, increasing organic matter and biodiversity, promoting carbon storage, and preventing erosion and desertification—in two words: more sustainable.

In addition to solutions that can be implemented through changing our land use patterns, we also need to pass more national legislation and a new water resource governance model. The current one was established 30 years ago when the impacts of climate change were not the absolute priority. Legislative approaches are ultimately limited in the face of global environmental changes, and for this reason, we must also continue to pursue international cooperative agreements. For the individual, tools such as the water footprint can enable the individual to understand the scale of impact that their choices can have in helping to restore and heal the local scars that globally induced water variability creates. The response to the current water challenges must be multi-dimensional.

The information presented in this essay should make us reflect on the urgent need to adopt strategies aimed at protecting water resources. From the single individual, to multiple readers, to large governance organizations, we should all adopt day-by-day strategies in our routines. Even the smallest acts can make a change, so we don't necessarily have to wait for legislation from above. Sometimes we can fall into despair given the negative impacts we witness, we can think that all is lost, but awareness and knowledge are the keys that allow us to improve and find solutions.

Voices of the Fellows
Water

Lucinda Coleman, Editor

Encounters with bodies of water invited individuals to embrace stillness: sensory engagement, a provocation for listening. Liquid languages gurgled and tossed as conversation, sometimes warmly comforting, sometimes bracingly cold.

Fellows carved through woodland tracks, drinking in leafy greens and rich browns: immersion in liquid forms dissolving body boundaries. Heightened sensitivity to constantly moving forms entranced and challenged resident artists, scientists, writers, dancers, architects, and filmmakers, boldly dancing on water while debating the form. The human body as a mere water droplet, in a community becomes a formidable and moving life force. We drink deeply, together, for we are made of water.

As a vessel, "water is not only held but holds. Water is an elemental deity, primordial substance, and sacred resource" (A. Rodrigues). If we can embrace water's mysterious life force, we can embrace the essence of human life as an ungraspable mystery and our efforts in representation can only suggest possibilities. And yet, as some residents discovered, "water is a model for our own adaptability, constancy, and care" (S. Maddox). If we begin with observation and with listening, we too may begin to cultivate deep sensitivity to the wisdom of the waters.

1 - Matthew Wong. *Taleggio Valley pools of water in 'Orridi'*, 2018. (Image courtesy of artist)

2 - Giulia Mattera. *Severing Tide*, 2018. (Image courtesy of artist)

ON LOCAL MEANINGFUL ENGAGEMENT WITH WATER CYCLES

Bring your awareness to the water cycle; informing oneself of the realities in your area. Where does your water come from, what journey has it been on, and where will it go next? Awareness and intentionality is the beginning of future action.
Ellen Avery, Dancer

Find pleasure in the systems around us. Build relationships with these physical environments, whether that means joining a group that cleans rivers, designing and planting bioswales to collect and clean runoff, or simply experiencing the waterways around you through recreation. Advocacy starts with empathy; empathy with relationships.
Samuel Maddox, Architect

3 - Tiffany Henschel. *Water field studies in Sottochiesa*, 2018. (Image courtesy of artist)

Be there, observe, and visit a body of water continually to participate in her process of transformation by actively looking, and allowing oneself to be inside the water cycle. And if this is done collectively, the effect is even larger. We have separated ourselves from the cycles of nature for too long. We need to be back inside the living web, and act according to the interconnections across living beings.
Luisa Brando, Architect

In Colorado we take the presence of water for granted, as it flows freely from our taps, even if it has been transported hundreds of miles through and across mountains to arrive at our homes. In other places, people carry it by the gallon in jugs from community wells. To be more efficient it is our responsibility to understand as much as possible about where our water comes from. As designers we can create installations, buildings, and landscapes that teach and show water cycles through direct experience.
Stacy Passmore, Landscape Architect

A person can engage their local community to initiate programs using water conservation on a personal and communal level. The main method should be community engagement to look after their environmental spaces. The long-term programs typically have a group of advocates to spearhead the programs on a daily basis.
Brian Yulo Ng, Water Treatment Business Lead

4 - NAHR workshop participants in the 'Life is H2O' video project, July 2022. (Image still courtesy of NAHR)

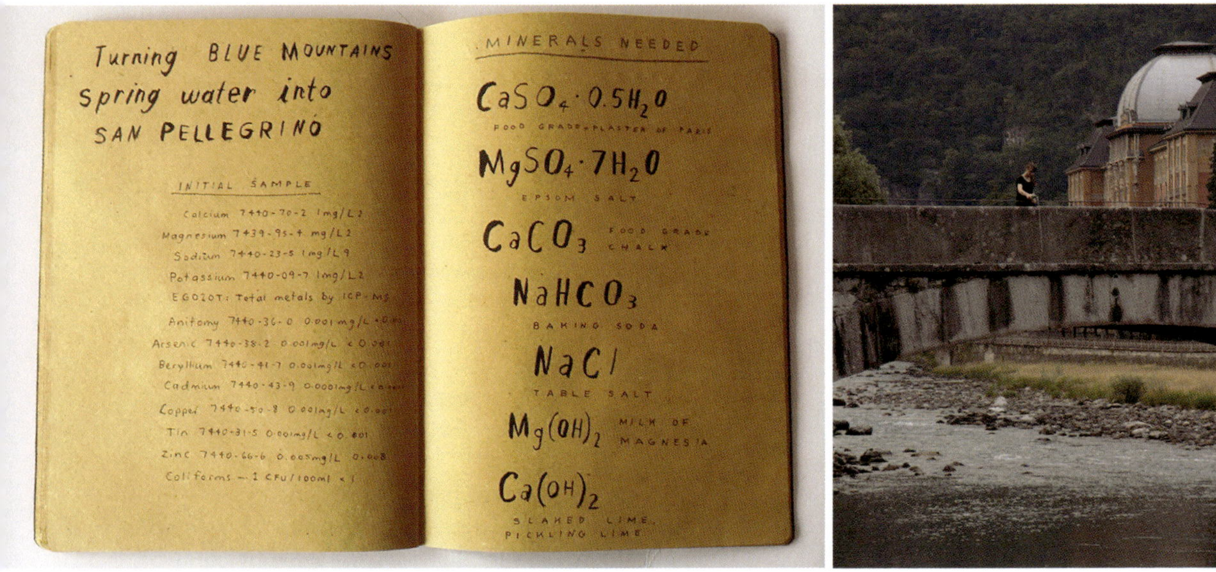

5 - Rachel Peachey and Paul Mosig. *A Convincing Likeness (sketchbook and explorations)*, 2018. (Images courtesy of artists)

ON COMMODIFICATION AND FUNCTIONAL VALUE OF WATER

We must create a mandate to focus on sustainability, replenishment and conservation of water to maintain natural features throughout the construction of new developments. From the initial design and planning of the masterplan, each development should look at the impact that their construction has on the surrounding environment. In order to ensure a long-term sustainable system, you must gain support from local government units, national government as well as private developers. For scalability and sustainability, programs need funding and support from the various interested groups as well as communities. It is only through the alignment of multiple groups that you can change long held practices.
Brian Yulo Ng, Water Treatment Business Lead

A narrative of therapeutic properties helped to alter San Pellegrino water from a fundamental part of a local ecosystem, to a commodity exported around the world at great profit. There are scientists creating a perfect replica of 'San Pellegrino' by adding the correct amount of minerals to tap water, thereby questioning the source of its reverence. We wanted to experience the water in its non-commercial state as part of its surrounding ecology and perhaps find some of its enduring magic.
Rachel Peachey and Paul Mosig, Artists

During my time with NAH*R*, I designed and crafted a series of puzzle-like water vessels that encouraged non-verbal interaction, cooperation, and play between visitors to the local fountains. Each vessel required two users to function correctly. The inspiration for these pieces came from regular meetings with recently arrived refugees in Vedeseta who were learning Italian yet struggling to interact meaningfully with their Italian neighbors.
Samuel Maddox, Architect

6 - Maddox's vessels in use at a fountain in Sottochiesa, 2018. (Images courtesy of I. Mazzoleni)

I have studied the role, design and function of the Italian *laac* seasonal livestock ponds that have been utilized for cattle grazing for hundreds of years by the migratory herders. Unique ecologies that have evolved in the human-built *laac* landscapes have led to small-scale but important associations between humans, animals, plants and water. As a result, not only the agricultural but also cultural and ecological systems now depend on them, yet they are threatened by climate change. This work led to several speculative proposals to insert new life into the *laac* to ensure they can be maintained, keeping their support of cattle, but also enhancing water storage through opportunities for recreation and human enjoyment. Ultimately, we learned through the project that these alpine landscapes and their hydrological systems that appear 'natural' are in fact the result of centuries of human curation and care.
Stacy Passmore, Landscape Architect

ON WOMEN AND WATER

Women's connection with water is primordial and enduring. Like women who bear the potentiality of children, water physically *holds* through the potentiality of the form in which it is contained. This vessel connection in holding literally takes shape and form through the spiritual axes of ritual, mythology, or religion, as much as it does through ongoing patriarchal instruments of systematic oppression. There is much to be learned in honoring the presence of history embedded in how current systems have come to be and are perpetuated. Across cultures and across time, we can witness both the imbued beauty and the weight of woman as water bearer, whether in relationship to the matrilineal labor of giving birth or the matrilineal labor of collecting water from the river or the well and walking miles upon miles. We can learn from these interconnections by reclaiming lived and living histories, which themselves can be found in women's own intimate relationships with water. These intimate relationships, whether individual or community, often speak to the embodied realm. By learning from them and by listening, we can seek to understand and create more earnest ecologies that are not removed from indigenous knowledge and cultural wisdom, nor the power systems that seek to invalidate or erase them.
Ashe Rodrigues, Dancer

7 - Stacy Passmore. *Laac Data Diagram*, 2018. (Image courtesy of artist)

8 - Ashe Rodrigues and Carley Culmo. *Wellsprings: A Subversion of Milk and Tears*, 2018. (Images courtesy of I. Mazzoleni)

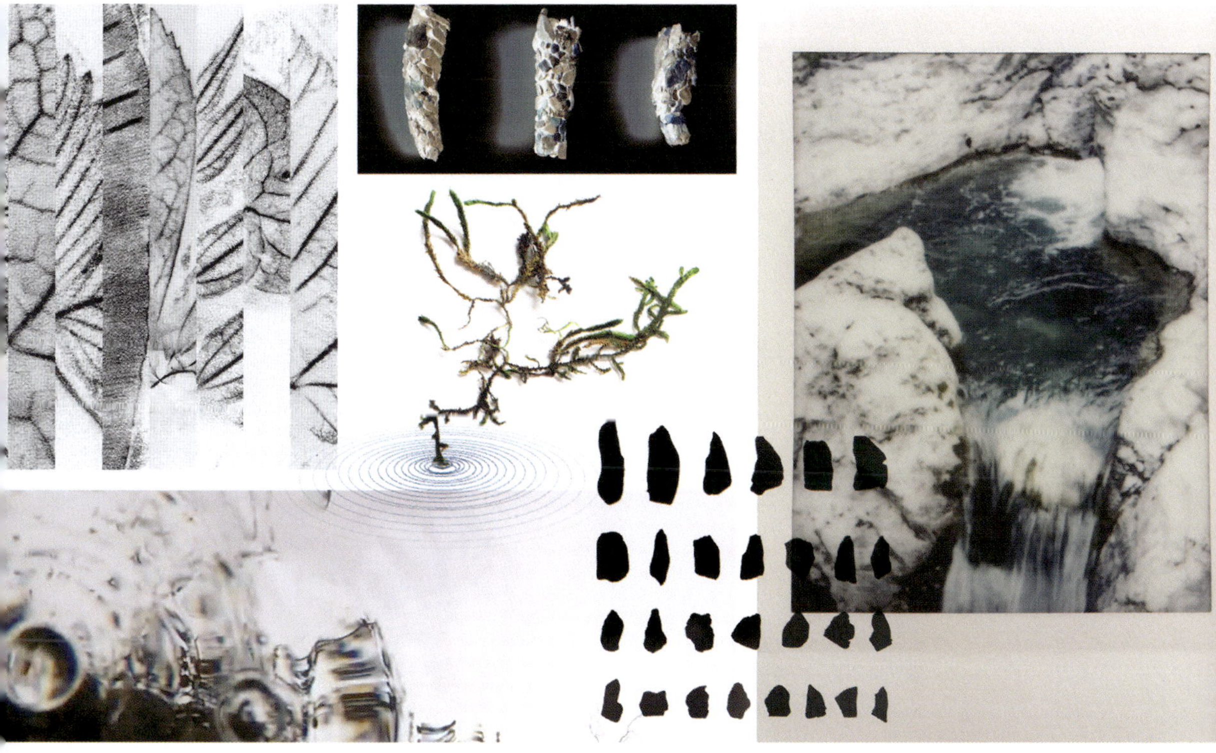

9 - Skye Morét. *Cues for Immersion*, 2018. (Image courtesy of artist)

Water is that which allows life to emerge. She is movement, she is process, she is cyclical. She is an element that connects, adapts, heals, and injures. She is rain as she is mist, she is a lake as she is a tear. Existent in plants, fish, humans, trees, mountains, and clouds, she is life.
Luisa Brando, Architect

ON ECOLOGICAL ENGAGEMENT

Art that works with the land, as in with reverence to it, will always have the capacity to inspire ecological restoration because it reveals the possibility in which we as humans can *remember* ecological relationships as a form of action. I think restoration is intrinsically linked with remembrance. Likewise, art as a form of expression that seeks to be seen, felt, and understood in some ways also seeks to be remembered when it is later perceived. Through remembering our human experience as an ecological experience, we have the potential not only to restore but to change, to develop upon, and to grow.
Ashe Rodrigues, Performance Artist

10 - Jill Anholt. *Acqua: Vita*, 2018. (Image courtesy of artist and P. Invernizzi, stonemason)

Water is, from a human perspective, a metaphorical and physical mirror of our bodies. It dialogues and reshapes (our) animal bodies, rocks, plants, lichens...Water, like thoughts, takes solid, liquid, and gas states. A direct material experience between physical body and water is the access to personal understanding.
Giulia Mattera, Performing Artist

11 - Ellen Avery in the Taleggio Valley, 2018. (Image courtesy of L. Coleman)

Many types of artistic work are valid contributions, but particularly those that are experiential and brokered through authentic interaction with a site. When one finds oneself immersed in the site, one experiences oneness with the environment, and in this moment feels *we*. We broaden our understanding of the ecological phenomenon in question, creating the potential for unique and specific responses.
Ellen Avery, Dancer

Whichever work makes us more conscious that we are all interconnected is key. The artistic work can be a written piece, a sculpture, a drawing, or a film. More than the art object itself, I rather think of the process of engagement of each person that creates, and how that process shapes the perception of the world and the relation with the more than human.
Luisa Brando, Architect

Work that renders complex systems more visible and tangible are particularly valuable—projects like *The Phenological Clock* (Jeremijenko et al., 2014) which empowers the viewer to see their environment as dynamic across time and species. Such simple shifts in perspective are the first steps toward more effective grassroots advocacy.
Samuel Maddox, Architect

In my work on the *laac* pond landscapes I used various interactive tools for engagement with the subject of water and, in particular, the measurement of change—both from the ground and from the aerial view. Because water and landscapes are dynamic, I'm interested in mediums that reveal those variations and show people both seasonal changes as well as those that are impacted by climate change. For example, the water levels of the *laac* ponds fluctuate seasonally, and so I created sculptural 'rulers', designed specifically for each pond, to indicate the estimated water volume and to show how the presence of water may change over time. By seeing the variation of the water, people may begin to understand how the water is both temporal and relies on us.
Stacy Passmore, Landscape Architect

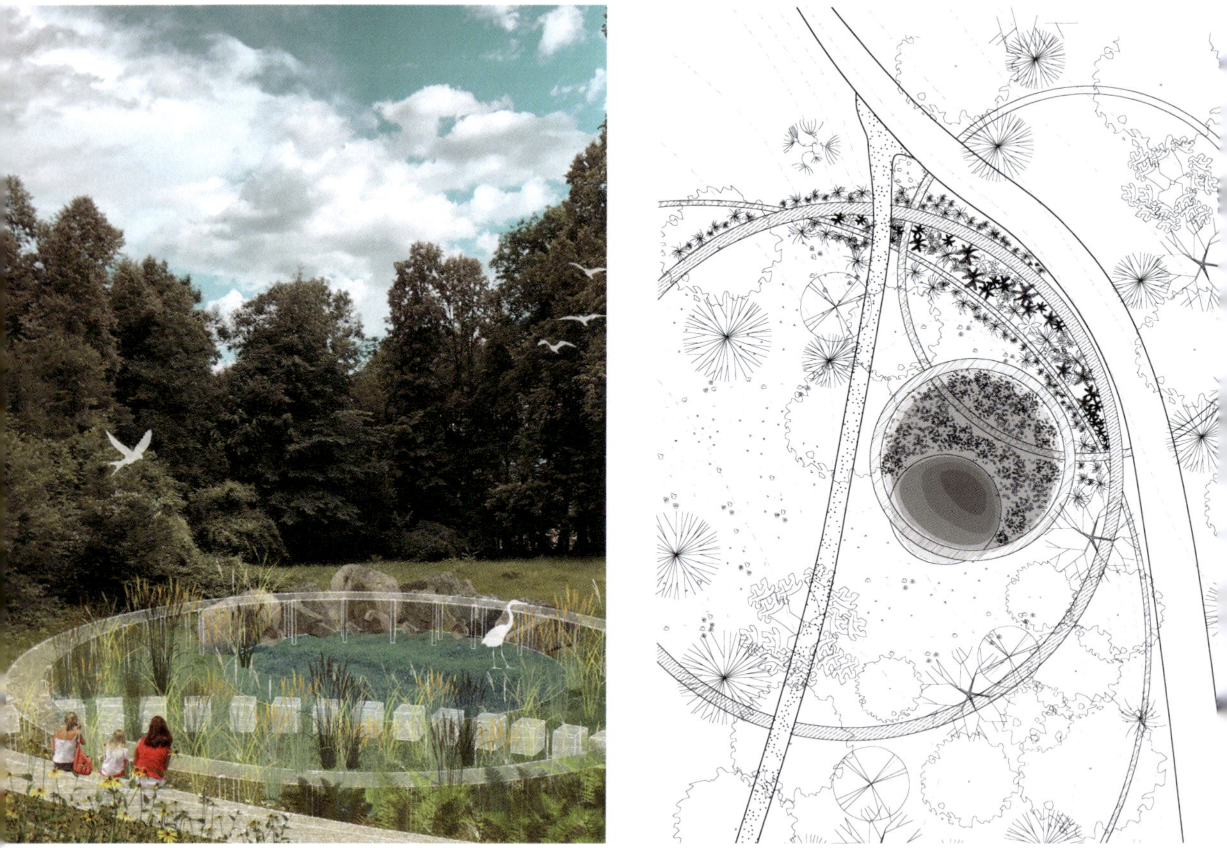

12 - Stacy Passmore. *Boardwalk - Taleggio Valley Laac*, 2018. (Image courtesy of artist)

13 - Luisa Brando, *An Aquatic Sense of Place-Carpenter Center Video Installation*, 2018. (Image still courtesy of artist)

This is an immersive installation on the Magdalena River, the main river in Colombia. It is an artwork that re-images and re-imagines Magdalena from a line (aerial map imaginary of rivers as static lines that have long been used to divide states and nations) to a space that connects the natural, the material and the immaterial.
Luisa Brando, Architect

Documentary films can address ecological issues with enough artistic or anecdotal merit to bring about a conversation. Films like the *Serengeti Rules* and *Seaspiracy* feature aspects of natural conservation, while still having a strong grounding in spectacle to draw a bigger audience to various causes.
Brian Yulo Ng, Water Treatment Business Head

14 - Brian Yulo Ng. *Animation Terrarium*, 2018. (Image still courtesy of artist)

15 - Rachel Peachey and Paul Mosig. *Enna river field studies*, 2018. (Image courtesy of artists)

We attempt to understand particular ecologies through direct observation and physical immersion. We use unconventional movement patterns to process particular ideas or uncover obscure connections. The documentation and sharing of this process can help to trigger surprising insights and pathways.
Rachel Peachey and Paul Mosig, Artists

I am drawn to ways of being in the world that do not center the human experience. If the boundaries of how we see ourselves go beyond the border of our personhood, we might care for these ecologies like we might hope to care for ourselves.
Rachel Peachey, Artist

It can be hard for artistic exchange to transcend everyday patterns. The closeness of our family provides a certain confidence to explore uncharted realms. We use a collaborative process where ideas, materials, and moments are passed somewhat like water through the relationship of Rachel, myself, and our two boys who are on their way to becoming something else.
Paul Mosig, Artist

[incomplete] codification of water by sensation

moonwater: magnets pulling in fever
puddle: sluice of water between toes
river: lungful of air after rain
lake: iridescence seeping
lagoon: crawling tails
channel: yeast left to rise
creck: murmuring before dawn
delta: bathwater draining away
stream: sound feathering onto skin
sea: velvet reaching for velvet
swamp: tide xylophoning against roots
glacier: disappearance of sun
ocean: shoulders of lover
ocean: baby teeth crunching beneath
ocean: throatful of life jackets

16 - Eunice Andrada. *[incomplete] codification of water by sensation*, 2018. (Poem courtesy of poet)

Grasses & Pastures

Imagining a Regenerative Economy

CHAPTER 4

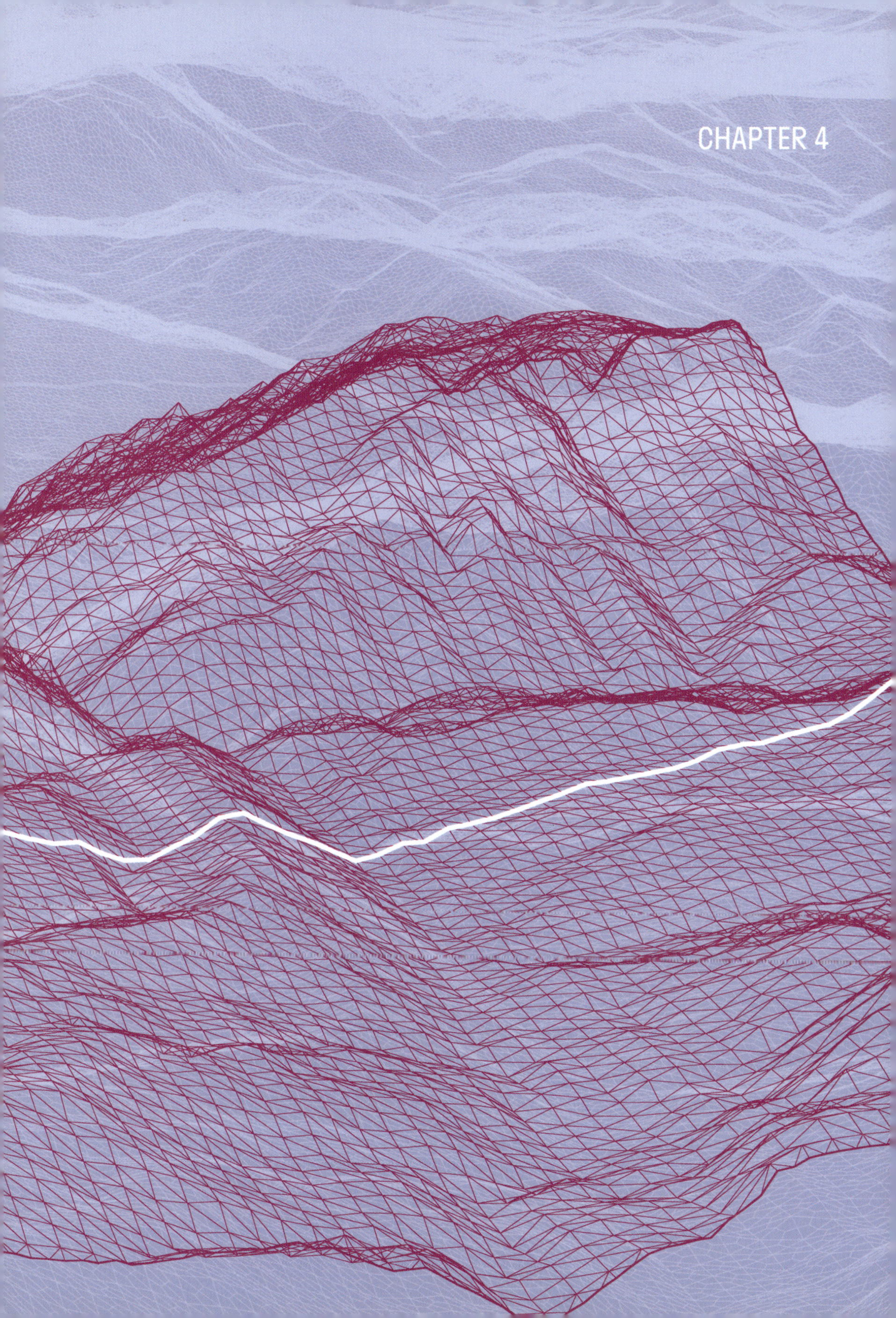

Introduction

Sara Invernizzi

Contemplating grasses and pastures stimulates various ecological, economical, and philosophical reflections, in particular when considering the place of the human among them. We have always had a vital relationship with grasses both as themselves but also as a conduit through which to engage with other organisms. From foraging, to hunting, and breeding, to cultivation, our dependence on grasses and pastures has always existed, albeit changing over time. We have generated a wealth of new landscapes where grasses now live in a reciprocal, interconnected exchange with human communities: for example, in the alpine areas below 1800 meters, meadows and pastures were obtained by the transformation of the pre-existing natural forests and ecosystems. However, for a long time the maintenance of the pastures, grasses, and soils, was a priority to farmers and dairy producers careful to not overexploit the delicate balance of these highlands. Today, delicate environments such as these, are at risk of disappearing, and their soils being degraded through misuse and global climate inconsistencies, along with the traditions of the farmers, or villagers, who take care of them. Our constantly evolving partnership with the grasses and pastures is also subjected to our global paradigm shifts.

Grasses are not only food, but also spontaneous and non-spontaneous architecture and design. They constitute a natural covering of the soil, which defends it from the atmospheric agents and from erosion. The specific characteristics of some types of grasses—in particular waterproofing, elasticity, and breathability—have made them important in some constructions, such as the creation of nests, as a base for animal dens, as bedding for animals in stables, as a building material for human architecture, inside walls, or as a cover for roofs. Furthermore, the human being has obtained from grasses the fibers for the packaging of clothes and objects.

Grasses not only dominate farmlands and rural landscapes for grazing animals, but they are also constantly present in our lives even in urban contexts as theorized by Gilles Clement as "third landscape", spaces abandoned by humans and gradually reconquered by different essences. Often these landscapes, from parks and nature reserves to large uninhabited areas of the planet, but also smaller and more widespread, almost invisible spaces, tell of a process of resilience that sees grasses as protagonists.

We shall see from the following two in-depth essays and the essay of collated voices that the articulated alliance between humans and the mountain grasses constitutes an exemplary case from which to study the circular regenerative economy. As you read these reflections, you are invited to recognize the critical ecological value of grasses that goes beyond sustaining life for humans and their extractive economic cycle. Grasses are extremely biodiverse providing a rich environment for an equally diverse web of insects, birds and other organisms; they have the fascinating ability to aggregate through the mycorrhizae present in the soil which enter into a relationship with the root systems, and create a dense network. At the same time, however, there are grasses that tend to overwhelm others, to "infest" the meadows, creating spontaneous monocultures, reducing biodiversity. In response to the difficulty of living in intercropping sites, some grasses have "specialized" and chosen extreme contexts, such as rocks, dry stone walls, and poor soils where competition with other species is decidedly less.

As Colin Sage writes in his essay, "Grass exists as a background for human and animal activity". Meadows and pastures are geographically varied from high altitudes to the extensive agriculture in hilly or flat terrains. We can think of the large expansive farms in the prairies of Argentina and Brazil, in large land holdings.[1] Today the cycle of production, distribution and consumption of cheeses and meats is far from being a circular economy. It is more of a hybrid model that combines the cyclical and traditional character of the 'harvesting' of pastures with the more modern linear model of production and distribution. While many of the elements for a fully circular business cycle are present, the current system appears to lack a focal point that would transform it into a fully circular and resilient economy. Sage's insights provide some possibilities to this urgently needed transformation in food production systems.

In Sabina Magliocco's essay we learn of the importance of the grasses, pastures and woods for the livelihood of those living off their local resources, and the interconnections between these environments and the survival of the humans that inhabit them. The author tells us "Two proverbs that capture this worldview are 'Il bosco non costa niente ma dà tutto' (the woods costs nothing, but gives everything) and 'Il bosco dà anche la minestra' (the woods also gives [us] soup)". Grasses, as part of the vegetable kingdom, are influenced by seasonal and climatic rhythms. All living creatures connected to the grasses and pastures have adapted to these rhythms: foraging, migrations, transhumance and related rituals refer to the close link between animal kingdom and plant kingdom. Magliocco's study into local cultures enables us to see clearly how the intertwining of human sacred and artistic cultural practices, with the diversity and cycles of the pastures, is key to a respectful and healthy relationship with these ecosystems.

[1] According to some estimates, extensive livestock grazing is practiced on 25% of the earth's surface (source: WWF).

In the third and final essay, we discover how intertwined our lives are with those of grass and pastures and with the other species that interact with them. As an architect engaged with the urban spaces and their current need for improved biodiversity, Floris van der Kleij writes, "To keep up the quality of life of all species, shared public spaces or 'commons' need strong conceptual ideas and attention: they must invite the deep experience of place". Further, artist Siobhan Rigg's discourse with the grasses in the high pastures hails them as a critical element which sustains life, and are worthy of their own constitution, as an element with its own independent agency.

Throughout the chapter, the reader will find grasses presented as constant renewal and in sensitivity to environmental factors. The authors show us the possibilities in rethinking our current linear economy into a regenerative ecologically attuned one, which does not forget the importance of all those elements that we are often led to forget, or to relegate to the cramped spaces of flower beds or in the high mountains. Creative practices like poetry, or other practices close to the land like foraging, help us to develop a language and culture that will improve our participation in the social and ecological commons of the grasses and pastures.

Grass, Food, and Climate in a Time of Uncertainty

Colin Sage

THE GLOBAL CHALLENGE

At a time when we are confronted by a global crisis of planetary heating and biodiversity loss, this essay provocatively suggests that 'grasses and pastures' may be presenting one of our most challenging and contested plant communities. Today, half of the world's habitable land is occupied by agriculture and 77 percent of this is used by livestock as pasture for grazing or for the growing of feed crops (Ritchie and Roser 2020). The expansion of agricultural land around the world is a consequence not only of a growth in human numbers but also a change in dietary patterns. This 'nutrition transition', as it is called, is characterized above all by increasing consumption of meat and dairy which has resulted in a huge growth in the numbers of animals reared—and

1 - Juan Barte. *Grasses of Val Taleggio*, 2017. (Image courtesy of artist)

slaughtered (an estimated 80 billion annually)—for their products. Increasingly, the question is being asked: if we are to restore some equilibrium with the planet, should we change our diet by replacing the large quantities of meat and dairy eaten by many in rich and middle-income countries, with more plant-based foods? If we were to do so, what would be the implications for farmers engaged in small to medium-scale livestock production and for the grasslands and pastures on which they depend?

Most intensive livestock production systems which have spread around the world are characterized by their scale in numbers of animals reared and their output volumes of meat, milk, and eggs. Known also as confined animal feeding operations (CAFOs) where most animals spend the majority of their lives indoors, pigs, poultry and a growing number of dairy cattle are fed a diet of protein-rich feedstuffs, such as soy, in order to speed growth to market weight and enhance yield, and are thus largely disconnected from their local resource base. However, most ruminants—cattle, sheep and goats—still spend at least part of their time feeding on grass or other browsing vegetation, though its composition, location, growth and role in animal nutrition varies very widely indeed around the world.

2 - Lucrezia Costa. *Air Breathing Us*, 2023. (Images courtesy of G. De Santis)

One of the key issues to emerge from recent research on the contribution of the global food system to climate breakdown is the disproportionate impact of cattle with regard to the release of methane, a key short-lived greenhouse gas, eighty-five times more powerful than carbon dioxide (CO_2) in retaining atmospheric heat over a 20-year span (Costa et al. 2021). While beef and dairy production score badly across a range of environmental variables per kg of output—including land and freshwater use, eutrophication, ammonia pollution etc.—their role in the generation of anthropogenic methane (i.e. arising from human activity, not natural sources) has fallen under the spotlight, particularly in countries where intensive cattle (and, to a lesser extent, sheep) are commercially significant. In the Republic of Ireland, for example, agriculture contributes 37 percent of the country's greenhouse gas emissions (the EU average for the sector is ten percent), most of which can be attributed to a national herd of 7.3 million dairy and beef cattle. Proponents of this agricultural strategy place emphasis on the grass-fed basis of the system, proclaiming its intrinsically 'green' and sustainable character given the 'naturalness' of grass. Critics, on the other hand, deploy a wide range of arguments that, while highlighting the climate burden of ruminants, also point to the wider costs of growing grass and extending pasture into uplands that would be more ecologically-suited to the restoration of native woodlands.

Grasses and pastures, then, are not without controversy—for example, there is a heated scientific debate about their relative capacity to sequester atmospheric carbon compared to woodlands—and can also be caught up in passionate discussions around the appearance of 'cultural landscapes' once forested now rendered as spaces for livestock grazing. Large scale habitat restoration is regarded by many terrestrial ecologists as an urgent strategy if we are to slow biodiversity loss. Indeed, there is a growing lobby that argues for the removal of agriculture from upland environments which should then undergo a process of 'rewilding' to enable not only the recovery of forests able to sequester carbon, but also endangered plant communities and even apex predators such as wolves. Such arguments present striking alternative scenarios that challenge and potentially threaten the existing basis of upland farming and herding communities such as those of the Alpine valleys.

In order that we make more space for nature it is likely that it is the less intensively farmed regions of the world where the focus of land use change will fall. Yet such marginal regions, even in rich countries, support—often precariously—the livelihoods of millions of people, with most dependent upon livestock to transform vegetation into commercial products. Indeed, so successful have these upland farmers been in carefully managing their resources and developing a body of tacit knowledge and skills passed down through generations that their production has even given rise to some iconic foods—particularly cheeses—highly regarded by connoisseurs. How might we reconcile, then, the urgent need to save space for nature, its biodiversity, and service to all living organisms, without compromising food output for a world of people, while supporting the livelihoods and cultures of upland human communities whose economies are closely tied to grasses and pasture?

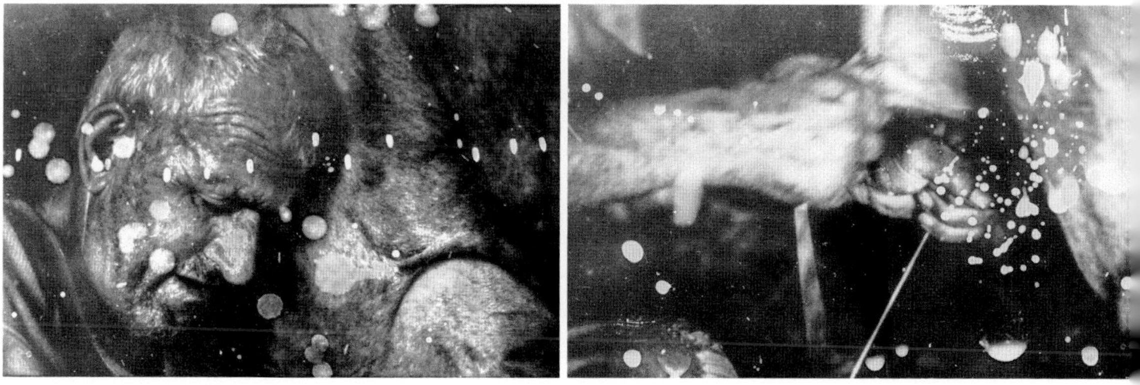

3 - An analog for a bio-inspired process. One of a kind copy on photographic paper. Juan Barte. *Truth, fulsome*, 2017. (Images courtesy of artist)

GRASS IN THE FOOD SYSTEM

Grasses are probably one of the least appreciated plant communities in popular imagination. Ubiquitous in comprising such diverse surfaces as suburban lawns, carefully manicured sports pitches (especially golf and bowling greens), to fields, meadows, rangelands and mountain pastures, grass exists as a backdrop to human and animal activity. For those of us with little botanical expertise, it is educational to learn of the enormous variety of grasses that are not only identified and cataloged, but which co-exist in plant communities on a spectrum reflecting their purpose and degree of human and animal intervention. Acknowledging that the first fields created for food production were reclaimed from woodlands, the opening up of space enabled light-loving grasses and herbs to flourish. Yet a landscape of herbaceous vegetation evolves, reflecting environmental conditions (climate, soil, topography) and its use. Early colonizing plants can gradually be suppressed by invasive trees and shrubs unless their emergence is controlled for example by grazing or fire.

4 - Grasses and pastures of the Italian Alps before the second cut, 2021. (Images courtesy of S. Invernizzi)

As agricultural science, and especially livestock husbandry, developed in Europe starting in the 17th century, greater attention was placed on different varieties of grasses to improve the diets of grazing animals, but also to restore nitrogen and organic matter to the soil. The inclusion of a grass ley, including clover and lucerne, as part of a crop rotation became a vital means to restore fertility—at least until the age of agricultural chemistry became ascendant in the late 19th century with its promotion of industrial amendments to the soil. The 20th century thus gradually saw the disappearance of a mixed farming model involving arable crops and livestock,where each farm produced multiple products within a largely circular economy of nutrient recycling which produced less environmental damage than modern intensive systems. Yet the pursuit of agricultural productivism as measured by output has been enshrined within the policy directives set by the European Union and national governments, and has led to specialization and intensification of single agricultural commodities which has become the 'norm' at farm level. Together with the major power brokers of the food system, operating in corporate boardrooms far removed from the land, the key imperative has been to drive down costs in order to produce food as cheaply as possible—irrespective of its nutritional benefits, health consequences or of the wider environmental externalities of production.[1]

For the farming community, the pressure to scale up has been intense as a consequence of declining margins under a cost-price squeeze. For dairy farmers this has invariably required them to increase the size of their herds. More cows—often substantially more—has huge implications for feeding and milking practices: does the farm have sufficient land on which to graze (and feed from cut hay or silage) its animals? Or must it purchase additional fodder and concentrates requiring greater financial expenditure? Milking technology, too, must adapt to reflect not only greater numbers of animals but the labor demands of undertaking this task once or twice per day. The installation of robotic milking or feeding systems—as witnessed on one Val Taleggio farm—may reduce labor inputs and cut the availability of jobs in the area, but requires huge investment ordinarily secured through bank loans, and leaves farms at risk of financial indebtedness if problems (animal disease or other unexpected hazards) should arise. It is not widely appreciated that rates of suicide amongst farmers are significantly higher than any other occupational category—3.5 times that of the general population in the United States—and while there may be a complex social pathology underpinning such personal and domestic tragedies, indebtedness is invariably the single most common factor. So while the agri-technology corporations promote their labor-saving innovations, we also recognise the capacity and determination of powerful interests to extract value from land without themselves being directly responsible for production.

When a visitor to a farm also observes large stacks of dextrose monohydrate—used as a sweetener to improve the palatability of a 'total mixed ration'—then the question arises as to the role of grasses and pasture in the feeding of animals producing milk for cheese. Although farms may be in scenic locations surrounded by grassy hillsides, we should not assume these continue to provide the sole means of sustenance for intensively farmed animals selectively bred for their high milk output. In this regard

[1] For further elaboration of this argument see Sage 2022.

appearances can be deceptive and we may need to be far more attentive to actual farming activities rather than to assume current practitioners are continuing an unbroken tradition of using grass.

It is understandable if we fall prey to such a delusion given that, in a European context, there is an incredibly rich cultural history around the agronomic management of pasture. From early medieval times farmers selected and sowed grass seeds that were designed to provide food for their animals both as fresh grazing but also as hay. From Pieter Bruegel (*The Haymaking*, 1565) through John Constable (*The Hay Wain*, 1821) to Camille Pissarro (*Haymaking, Éragny*, 1887) and beyond, European artists have captured the vital moment of 'saving the hay', constructing vernacular stooks and stacks, and ensuring that their animals would survive over the winter in a temperate, even hostile, climate. The often communal nature of this task, when neighbors would work for each other within a spirit of reciprocity and conviviality, was vital to their own survival as a community dependent upon stored roots, coarse grains and, of course, fresh, seasonal, and fermented milk products.

5 - The storage of hay during the mountain pasture season: Alpe Sella in Val Taleggio (BG), 2021. (Images courtesy of S. Invernizzi)

Grasses and pasture, then, have long played a critical role in farming communities with their importance rising with altitude and latitude. While pastoralism displays many different forms reflecting complex geographies, the movement of animals in accordance with seasons has, like haymaking, a rich cultural history across Europe.

The practice of transhumance to make use of mountain pastures during the summer months has enabled livestock owners to produce a "gold of the uplands" in Grasseni's words, which are cheeses that capture the floristic diversity, aroma and taste of these mountain meadows. The absence of animals from lowland villages also allowed those more intensively grazed fields time to recover and to yield a crop or two of hay before the flocks and herds returned. The practice of *alpeggio* has been an important ritual throughout the Alpine region demonstrating how rural communities secured their food through ecological adaptation and social cooperation. That they also produced foods of distinction has enabled many to flourish given a wider gastronomic interest in their products.

6 - Pasture practice at 1700 meters above sea level at Piani di Artavaggio, Orobie Prealps, 2020. (Image courtesy of S. Invernizzi)

Yet this is an uncertain time for such a production system, dependent as it is on the continued support of consumers willing to pay premium prices for geographically designated products while the urgency of action on climate is creating a powerful narrative around plant-based alternatives. It is increasingly apparent that dietary change is an indispensable tool in a complex portfolio of civil society responses to climate breakdown, and the development of alternative proteins including from 'precision fermentation' technology may well 'eat into' (pun intended) the market for livestock products. Indeed, many of the major meat processing companies have already rebranded themselves as leaders of the new 'protein industry' thereby repositioning themselves as 'planetary saviors'. Plant-based milks, too, have grown their market share and increasing numbers of eaters—especially younger generations—identify as vegan or vegetarian.

For livestock farmers such developments might be disregarded as a transient fickleness of the market. Yet amongst policy-makers willing to listen to the science, regulations are beginning to take shape that will inevitably, and arguably belatedly, require reductions in methane emissions from ruminant animals. How will such restrictions be applied? For intensive sheep and cattle farms in lowland environments, options in arable production may be available and ought to be encouraged where practicable in order to produce more plant-based food. For upland farmers, however, with smaller herds or flocks practicing traditional and more ecologically suitable forms of production, effort must be made to ensure a 'just transition' whereby their livelihoods can be ensured with the help of wider social and economic networks of support that value their continued existence.

GRASSES AND PASTURE IN THE CLIMATE TRANSITION

Cheap, highly processed food has become the hallmark of our era. It is the product of a deeply flawed logic that has come to regard food as a source of profit and wealth accumulation rather than as the foundation of human well-being, social vitality and cultural significance. For most of us as eaters in rich and middle-income countries, we are presented with the greatest range of products requiring the smallest expenditure of household income in history. Yet the paradox of the contemporary food system is that around one billion people are chronically food insecure, almost two billion are overweight or obese, and at least 11 million people die every year from specifically diet-related illnesses (GLOPAN 2020). Simply regarded as 'consumers', the food

7 - Semi-intensive farming in the Lodi area (Po Valley): the grasses disappear behind the plastic, the manure behind reinforced concrete walls, 2020. (Images courtesy of S. Invernizzi)

8 - Ancient water mill in Lirone, Italian Alps, 2019. (Image courtesy of S. Invernizzi)

system with its ubiquitous availability of processed products has infantilized and disempowered a general population lulled into a narrative of convenience and low price. In short, the contemporary food system has become decoupled from parameters such as the numbers of people fed healthily and sustainably.

However, over approximately the past three decades, increasing numbers of people have sought to engage with a different vision of food; one that is more localized, preferably organic, and potentially healthier. This emerging coalition of diverse actors concerned about their food and how it is produced now constitutes an important bloc in civil society that arguably has moved beyond the mere purchase of 'better products'. Rather, significant numbers of people are expressing opposition to the status quo, demonstrating that alternatives are not only practically feasible, but also deliver a host of other co-benefits, including ecological regeneration, community building and improved well-being (Sage et al. 2021). The unfolding climate catastrophes of recent years are beginning to sharpen public attention even while political leaders are distracted by maintaining economic growth at all costs. While there are disparate and discordant voices within civil society, there is evidence to suggest the foundations for an agenda of social change is emerging. Some of its key features are being articulated most forcefully by a younger generation ('Millennials', 'Gen Z') many of whom comprehend the devastating consequences of continuing to pursue 'business as usual'. The challenge now is to harness this energy and courage with the tools that we know provide for an effective alternative pathway of transition capable of delivering food security, social justice, better human health and planetary ecological equilibrium.

Arguably, the single biggest problem that we must overcome is to move away from the atomized model of consumption and towards a more aggregated and collective response. The market works best by individualizing us as consumers despite the needs and aspirations we have in common. Recognizing the ways in which food opens up spaces for conviviality—through growing, purchasing, cooking and eating—creates opportunities for redesigning many conventional practices at the neighborhood level. The proliferation of community gardens, the expansion of solidarity purchase schemes (e.g., GAS in Italy) and a growing consciousness around a range of ethical considerations (animal welfare, the rights of nature, climate breakdown), demonstrates the existence and growth potential of this emerging movement.[2]

There are, however, several important considerations before we can embark upon a program of change. First, we have to recognize that we now live in a world where silo thinking is not only redundant, it is dangerous to the complexities of our age. We have to be able to address multiple policy fields simultaneously, synergistically, and where there can be little room for trading off one issue (economic growth) for another (social justice). Secondly, we can no longer afford short-term thinking allowing events to unfold as they will ('trust in the market') without some coherent vision in place to guide us. Scenario planning is thus a vital tool in the repertoire, not only in establishing a road map, but in ensuring the widest possible consultation in its preparation. This leads into a third indispensable element: that of building collaborative initiatives to reshape production and consumption norms. When we consider the extraordinary abundance of ingenuity and expertise which exists in our societies, it is simply unfathomable to comprehend how we are in this current predicament. Yet we have created disciplinary walls between academics, whose knowledge is set apart from most of civil society that in turn is disregarded by business (except in their role as customers). Consequently, we have to find new arrangements capable of bringing together a diverse selection of professionals, activists, business-people and artists capable

9 - A tool pivotal for human productivity, reclaimed by the grasses. Beth Robertson. *Fraggio*, 2023. (Image courtesy of artist)

[2] Indeed, one might argue that food has become increasingly politicized not only on the global stage, where grain shipments represent life or death for the poorest and most food insecure regions, but at sub-national level in rich countries where citizens seek to achieve alternative visions for more localized food systems. Often experimental in design many projects fail, but those that persist can significantly revitalize thinking around food, not only calling into question current arrangements—how food is produced, by whom and at what cost to the environment—but through devising collective responses including efforts to enroll and mobilize local political authorities. The growth of food policy councils is one such development, while solidarity purchase schemes (GAS in Italy) have proven a highly successful mechanism for (re-)connecting consumers and producers. For more, see Sage et al. (2021), Giambartolomei et al. (2021).

10 - Danica Liongson. *Research process*, 2022. (Images courtesy of artist)

of initiating new conversations around the food system as well as its wider place in the world. The question, then, is how might such initiatives have any connection with farmers managing grasses and pastures in upland environments?

If we are to have any hope of addressing the existential threat posed by climate breakdown, then ensuring a just transition will be vital. That is to say, the wholesale imposition of strict regulations on a sector—indeed on a way of life, such as livestock farming—is unlikely to be successful if this results in the demonizing of certain foods. Yet 'business as usual' is also not an option. Farmers will be required to change, to accommodate to new targets that might require a level of destocking, that might involve alterations to existing land use such as afforestation (for carbon capture, for habitat regeneration), and they will need to be compensated for such changes. Clearly, there is considerable scope here for innovative financial instruments overseen by public policy that better valorize ecological services and the stewardship of farmers. But it also opens up opportunities for new partnerships and collaborations: with outsiders who validate, encourage and support their efforts and perhaps can add value to the measures that farmers undertake. While this might involve ensuring a market for existing products—through solidarity purchase or community supported agriculture—the portfolio of products themselves might change as farmers and producers develop more agro-ecological measures for which premiums may apply. In this respect, building a more specifically territorial and landscape approach could become a central tool to ensuring that farm livelihoods are ensured even while production volumes may decline. The key is that designations of quality, in an increasingly differentiated market, wins over output by volume of singular brands. This represents something of a paradigm change to the existing food system where 'piling it high and selling it cheap' has generally prevailed. Can upland farming communities that have created livelihoods for many generations by managing grasses and pasture and their grazing livestock adapt to these new circumstances and lead on this necessary transition?

CONCLUSION

This essay has presented a critical perspective on the current food system, one where neither upstream producers or downstream 'consumers' are the major beneficiaries. 'Taking back the middle' has become a refrain in certain circles, where the scale of corporate concentration and the drive of productivism is seen as a major obstacle to ensuring that the entire world eats well and within planetary boundaries. Dietary change in the rich and in many middle-income countries will be necessary if we are to meet those goals. The human and environmental health burden of intensive livestock production is simply unsustainable and threatens to overwhelm ever-narrowing carbon budgets and health services. Planetary saviors promoting alternative proteins may suggest that death-free meat eating can continue, while others extol confined animal feeding operations utilizing rations that might comprise a diminishing proportion of grass in favor of concentrates and methane-reducing seaweed. If the food industry continues along this high-tech path of AI, robotics and cellular agriculture then we look toward a future that not only excludes eaters as stakeholders, but likely renders grasses and pasture as an archaic, analog form of animal feeding to be eliminated. This is the vision that proponents of precision fermentation offer: large-scale industrial manufacturing of protein that renders livestock as entirely superfluous to food supply and thereby releasing land for rewilding, species recovery and carbon sequestration.

11 - Ming Lin and Heidi Maribut. *Radio for Rural Futurities*, 2019. (Images courtesy of artists)

On the other hand, a broad constituency of consumer citizens concerned not only about the quality of their food, but how it is produced, by whom and where, and working to ensure that production systems are transparent and demonstrate high ethical standards could transform food systems worldwide. The key here is to insist that consumer citizens take greater responsibility for their food purchases, not simply by their shopping choices or reducing kitchen waste, but by reaching out to others including those back up the food chain. Building strong and durable alliances with farmers is critical here, but will require patience and tenacity given the fierce independence that the agricultural community traditionally demonstrates. While it may be reluctant to relinquish complete control over its own affairs, farmers should be encouraged to appreciate the opportunities, and potential rewards, that democratizing the food system presents. Building closer relationships with the consumers of their products provides a social validation of their work, and creates sources of income that do not pass through the hands of intermediaries.

Reterritorializing food systems is a first step in the new agri-food transition, in which natural geographical boundaries may define a regional foodshed. Without imposing fixed spatial parameters, each region will emerge reflecting its own historical, socio-cultural, and ecological circumstances, and an initial task is to appreciate the particular agricultural and production characteristics of various territories. This is not to create autarchic regions engaged in the pursuit of self-sufficiency, but to better understand what can be provided to ensure the basis of a healthy, sustainable and culturally appropriate diet in the region. The current food system is extractive and linear with produce being transported over long distances; reterritorializing food supply provides an opportunity for more circular, closed-loop arrangements, reducing waste and other inefficiencies and valuing foods, ecological services, labor and other resources, including grasses and pastures.

In this respect the husbandry of animals on natural vegetation at stocking levels that do not exceed clear agro-ecological parameters—and not just regrowth of pasture but methane emissions that are balanced by sequestration in soils and plants—can form a vital part of sustainable food systems. Meat and dairy products therefore have a future, although at substantially reduced volumes of consumption in rich countries. The prospect that alternative proteins will displace animal tissue is unclear at this stage. Alternative protein products may simply provide an additional rather than substitute source of protein. For the dairy sector there are challenges too, although the fractionation of liquid milk into a wide range of food and industrial products (including infant and child formula, sports nutrition, the beverage sector) makes this food sector less immediately threatened by plant-based alternatives. For both mainstream milk and dairy, cutting costs in pursuit of profit may well have consequences for animal feeding regimes with implications for pasture. But for quality cheese makers and their milk suppliers, the role of well-managed and diverse grasslands will remain critical to their business. Nevertheless, in the interests of remaining vigilant about the future, we should probably learn to appreciate more the formidable roles played by grasses and pastures, especially their wider social, cultural, and ecological dimensions.

Lessons of the Fields and Pastures: Traditional Environmental Knowledge and Reciprocal Relationships with Nature in the Italian Prealps

Sabina Magliocco

When thinking about pastures, meadows, fields, and the plants that make up these ecosystems, we must also consider the folk knowledge, traditions, customs, and rituals associated with them that link rural peoples to their environment. Westerners often assume that only Indigenous peoples know how to live on the land in a sustainable manner supported by a spirituality linked closely to nature. Yet this way of living was once practiced in many parts of Europe as well. Anthropologists studying Indigenous cultures have observed that when traditional peoples live in a sustainable manner, they do so because of deep knowledge of their environment combined with emotions and ideas encoded in their cosmology (Anderson 1996, p. 56). Since we are currently living in a state of climate crisis created by the Western capitalist extraction of natural resources for sale to a global market, it is hard to imagine that Western Europeans once participated in cultures that had similar relationships to nature and the environment. These modes of subsistence and the worldview that sustained them continued to be practiced as recently as a generation ago in Europe's 'inner areas': regions that remain relatively isolated from urban centers, and are linked to pastoral and agricultural economies. One of these is the Prealpine region in the province of Bergamo, specifically in the Val Taleggio, where inhabitants combined high altitude bovine pastoralism with foraging in the rich woodlands, meadows and pastures to make a living in this precarious ecosystem.

What lessons can the lifeways of the Val Taleggio and other peripheral areas teach us today, as we search for new pathways for engaging with the environment—models that are based not on the extraction of natural resources, but on responsible stewardship and reciprocity? And how can pre-industrial lifeways be adapted for a post-industrial future? To answer these questions we will explore two contrasting communities: the inhabitants of the Val Taleggio who either remember traditional ecological knowledge and practices, or are actively reclaiming them; and small urban groups in the wider region of Val Brembana (one of the two main valleys in the province of Bergamo) who embrace the 'new animisms', seeking to revive, reclaim, and reconnect with traditional methods of living within the natural environment. Both groups can teach us important lessons about relating to grasslands and pastures, as we strive to create a better future that will not rely on extractive practices and cultivate a reciprocal relationship with the environment.[1]

[1] This essay is based on research conducted in the Val Taleggio in 2016, as well as information published by other ethnographers working in the area (Grasseni 2017), and, in the case of new animists in Bergamo, by ongoing digital contact with the community. Research methods included oral history interviews, archival research, and participant observation.

1 - Juan Barte. *Turning the hay*, 2017. (Image courtesy of artist)

"IL BOSCO NON COSTA NIENTE MA DÀ TUTTO": THE WOODS AND PASTURES AS A SOURCE OF SUSTENANCE

Before the advent of advanced consumer capitalism following the Second World War (1938-1945), the inhabitants of the Val Taleggio were similar to most rural Italians in their reliance on subsistence farming and pastoralism. Residents of the Val Taleggio raised milk cattle—the 'Bruna', a breed of dairy cow uniquely adapted to life at high altitude, similar to the Swiss Brown cattle. They supplemented pastoralism with small-scale farming and gardening, especially on the western end of the valley, around Vedeseta, and foraging. The subsistence economy entailed a strict dependence on nature and its cycles and processes; as a result, most people had a high degree of traditional ecological knowledge (TEK) that governed all household activities, from food procurement and production and the raising of livestock to craft production. This knowledge was handed down mostly through observation and imitation. Local tradition includes a proverb: *'Non aspettare che t'insegni il lavoro; lo devi rubare'* (don't wait for me to teach you to work; you must steal it), which meant that young people were expected to learn tasks by observing their elders, imitating them, and taking initiative to undertake similar work, without any formal verbal instruction. They were also encouraged to observe multiple people doing the same work, so they could 'steal' (i.e. imitate) work techniques, again without having them verbally explained. Asking questions or asking for verbal instruction was discouraged; those who did got scornful responses. This approach built initiative and strong observational skills among young people, which served them well as they grew up in a challenging environment.

For small-scale herders and farmers living on the land in precarious economic conditions, fields, pastures, and woodlands provided important resources for food, craft production, healing, and additional income. Spaces outside the inhabited area of the town were often referred to simply as *il bosco* (the woods), even when these spaces comprised a variety of environments, from wooded areas to fields, hedgerows, and pastures. For this reason it is impossible to talk about the cultural relationship to grasses, fields, and pastures without also touching on the woods. These areas offered a rich source of nourishment, both animal and vegetable. Two proverbs that capture this worldview are *'Il bosco non costa niente ma dà tutto'* (the woods cost nothing, but give everything) and *'Il bosco da anche la*

2 - Agapito Locatelli, a farmer from Val Taleggio, shows the edible plant *Persicaria bistorta*, good for cows and people and called 'tortorèla' in the local dialect, 2019. (Image courtesy of S. Invernizzi)

minestra' (the woods also give [us] soup). Birds, small mammals and deer of various species were hunted for meat. Walnut and hazelnut trees, which often grew on the edges of fields and pastures, provided nutmeats, an additional important source of protein. Wild strawberries, raspberries, and blackberries were gathered from the hedgerows during the late summer to eat as fruit or to preserve as jams and jellies; the hips of wild roses, rich in vitamin C, were also used to make jelly. A variety of wild greens were gathered in early spring, and cooked like spinach or used in sauces for pasta and gnocchi; these added fiber, vitamins and minerals to the diet.

3 - Wild herbs and grasses were eaten raw as salad or used for cooking, including Good King Henry (*Blitus bonus henricus*), called cuöei or cugoi; selene (*Silene vulgaris*), called 'cornagì'; dandelion (*Taraxacum*); nettle (*Urtica dioica*); goat's beard (*Tragoprogon pratensis*), or 'barba de bec'; cat's ear (*Hypochaeris uniflora*), or 'costolina', 2020 (Images courtesy of I. Mazzoleni, M. Macconi, and S. Invernizzi)

Before the advent of modern medicine, villagers turned to the pastures, woods and fields for herbs with healing properties that could help with everyday ailments. Interviewees reported gathering and using a number of plants for medicinal uses in the period before World War II and the diffusion of commercial over-the-counter remedies for common ailments. For colds, a tea (decoction) was made from linden and/or elder flowers and given to the patient to drink. The patient's feet were rubbed with olive oil and propped up near the hot stove to keep warm. For chest colds and coughs, flax seeds were ground and made into a poultice with *sugna*, a type of pork fat. This was placed hot on the patient's chest, on a clean cotton or linen towel; the heat was thought to help loosen congestion. Chamomile tea was the remedy of choice for headaches, while for bladder infections, 'gramigna' (*Cynodon dactylon*) was boiled and drunk as a tea. Mallow could both be used in a poultice to soothe a toothache, or drunk for its calming effect.

The fields also provided remedies for the garden. In addition to yielding a valuable wood for furniture-making, the walnut tree produced hulls around its fruit that farmers macerated in water to create a powerful spray against weeds and insect pests.

An important principle which applied to gathering greens and herbs from the forest was never to take the entire plant. Only the usable leaves were removed; the rest of the plant and its roots were left intact so the plant could continue to grow and reproduce. The plant could grow new leaves for later cutting, and the technique allowed new plants to grow in the same location for gathering the following year. This traditional ecological knowledge ensured a sustainable supply of greens and healing herbs year after year.

Villagers could earn additional income by foraging plants and selling them to merchants, either locally or in the nearby cities of Milan and Bergamo, who traded in them. The Biava Salvioni family in Sottochiesa owned a compounding pharmacy, and as late as the 1930s, young women from the area earned extra income for their families by collecting healing herbs from the woods and selling them to the pharmacists. The Vitari sisters, originally from Regetto, reported collecting the following herbs; however, they did not prepare them themselves, nor did they know their precise mode of use; they collected them to sell to the Biava Salvioni pharmacy.

Herbs had to be gathered at the right time, when the part that was used in medicine was ripe or at its most potent. That meant the young women who did this work were familiar with the plant's life cycle and its relationship to the broader seasonal cycle. Healing herbs gathered at the wrong time were useless to the pharmacist and would not bring any income.

In all these gathering practices, interlocutors emphasized the principles of sustainable harvesting. The point was never to raze the forest or despoil the fields, but to take only in proportion to one's need, always thinking of future needs. In this way, the ethos at the center of the relationship between valley dwellers and nature was similar to the one described by Potawotami ecologist Robin Wall Kimmerer as characteristic of Indigenous North American cultures (Kimmerer 2013). Rather than thinking of sustainable principles as belonging only to certain cultures, it would be more useful to think of these principles as characteristic of most subsistence economies.

4 - Some of the herbs the Vitari sisters gathered included Gentian (*Gentiana*), Rue (*Ruta graveolans*), Juniper berries (*Juniperus communis*). Ornamental plants such as Juniper (*Juniperus communis*), Mistletoe (*Viscum album*), and Cyclamen (*Cyclamen repandus* or *C. purpurascens*) were gathered from the woods and sold to merchants during the holidays for use as decorations in the cities. Local villagers gathered moss to use in their *presepio* (nativity), but until the 1960s and 1970s, few villagers decorated their homes during the Christmas season. Images from 2017 to 2023. (Images courtesy of M. Macconi and S. Invernizzi)

NATURE AS A PLACE OF ENCOUNTERS WITH THE SACRED

The woods, fields, and pastures were not just sources of sustenance for the inhabitants of Prealpine areas in Italy; they also had symbolic value through their association with mystery and the sacred. Many customs and traditions recognized the animate qualities of plants, fields, and natural features, and involved rituals to propitiate them, bringing good luck and plentiful harvests. While some of these rituals existed under

5 - NAHR Fellows observing the summer solstice by interacting with traditional grasses and herbs, 2016. (Image courtesy of I. Mazzoleni)

the auspices of the Catholic church, others were part of informal, unofficial culture, preserved through oral tradition. Both types of custom perpetuate patterns of relating to nature. These are likely deeply rooted in an Italian history that predates the imposition of Christianity on the indigenous, pagan religions that did not separate humans from nature to the extent that monotheistic religions do. Many of these customs were associated with spring's arrival, an important time in the yearly cycle when weather could mean the difference between a plentiful harvest and a winter of hunger.

Throughout the Prealpine region of Italy, the custom of *rogazioni* (rogations) marked the arrival of spring. These consisted of prayers, songs, processions, and acts of personal penance intended to draw divine blessings on the fields, pastures, springs, and the fruits of the earth. Some scholars trace the origins of the rogations to the pre-Christian tradition of the Ambarvalia, a similar custom in which the ancient Romans propitiated the deities for a good harvest. The major rogations took place on April 25, while the minor ones coincided with the three days preceding the Feast of the Ascension, generally celebrated about six weeks after Easter, between April 30 and June 3. At these times, villagers would exchange colored eggs and flower crowns as symbols of new life. In preceding weeks, women and girls would gather flowers and herbs that were used to dye eggs and weave crowns. On rogation day, villagers would form a procession that would begin at the church and wind its way around the entire territory of the town, as participants prayed and sang songs of thanksgiving for a good season and a bountiful harvest. In some towns, the processions would go on until sundown, culminating in a mass at the village church. Many Alpine and Prealpine towns are reviving this custom as an occasion to affirm the bonds of community while celebrating the arrival of spring (Lombardini 1926, 2002).

6 - Chestnut bark horn, a musical instrument that is played by mouth made of chestnut bark during the spring in the Lombardy Prealps. It is an ephemeral instrument, as it tends to dry up quickly, losing the ability to emit sounds, 2023. (Image courtesy of M. Macconi)

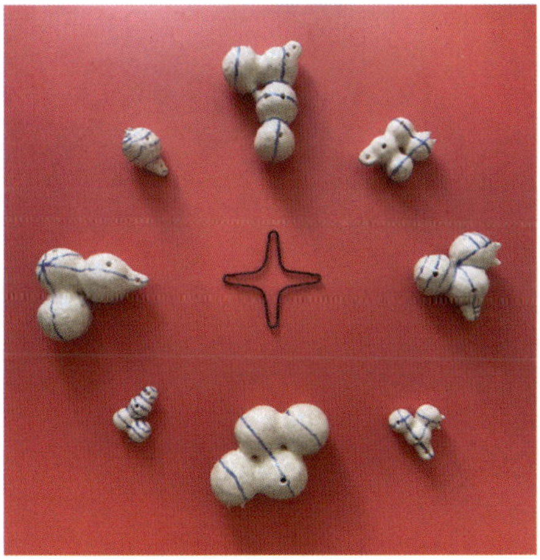

7 - Heloise Tunstall-Behrens. *Atmospheres*, 2023. (Image courtesy of musician)

A less formal custom whose roots also go back to pre-Christian practices is known by several names, including *ciamà l'erba* (calling the herbs), *sunà da mars* (playing music for March), *ciamàda de mars* (calling of the March), and *festa dei sampugn* (feast of the bagpipes). This custom took place at the beginning of March, when groups of young men, called *marziröi* (March boys), would run out into the fields with bagpipes and improvised instruments, making as much noise as possible to awaken the plants that would provide precious pasture for livestock and food for villages who had endured a long winter without access to fresh vegetables. In some towns, the young men painted their faces and dressed in disguise as they went door to door collecting offerings, which were later used for a feast in which the village children and youth participated (Lombardini 1926, 2002; Tazzoli 1935). Young people would bring their intended partners to the feast, and those who did not have one would take the occasion to declare their intentions to their beloved. The symbolic association of the return of spring with the youngest members of society—the adolescents who represented the source of the culture's renewal as they formed couples and eventually new families—demonstrates how deeply entwined relationships with nature were with social relationships. Plentiful herbs meant a good harvest, plentiful milk from well-fed cows, enough food to last another winter, and abundant resources so that life itself could continue: the towns and villages of the Prealpine valleys could thrive.

These traditions illustrate that in pre-industrial Prealpine Italy, the natural world was regarded as animated, sacred, and responsive to human participation. The calendar was marked by rituals in which villagers interacted directly with the fields and pastures, treating them as alive, sentient, and connected to a participatory system linked to both formal religion and informal social practices that were necessary to ensure a successful harvest. This ever-renewing cyclical process was itself regarded as sacred. For rural Italians, the sacred was part of the natural world, and the natural world with its processes was sacred.

This association of nature with the sacred was rediscovered by the Romantic movement in the 19th century, and is an important source of the reclamation of TEK and nature more generally by 20th and 21st century environmentalists, neopagans, and the 'new' animists.

Calling spring

«Èrba föra che l'é maarz»

At the extreme limit of winter,
Uro-woman marks the passage
between the ancient time,
the time that is - of abandonment -
and the time that will be.
Old puppets of stale straws
they dissolve in smoke.
Bursts of resinous essences
in the heat of the bonfire.
Go to the fields playing cowbells.
To go for a collection, to evening:
Green young collect offers
Green herbs and soft soil.
In the red glimmer of the embers
mask wearer
it doesn't know, but it might know
the time that will be.

While I was looking for a skull to hang on the lintel of my imaginary Carden, in an abandoned maggengo, I found a mask of uro: it was a twisted root of stone pine.
Even evil spirits are afraid of the terrible masks and macabre trophies hanging from the doors.

8 - *Calling Spring*: photo and text related to the spring rituals called 'calling spring' in some areas of the Alps, 2022. (Image courtesy of S. Invernizzi)

NATURE AS A SOURCE OF INSPIRATION

By the mid-20th century, life in the Prealpine valleys of Italy had changed radically due to the increasing influence of industrialization and the global market. Young people left smaller towns and villages for jobs in the industrial centers, resulting in the depopulation of some communities. People responded to the end of the traditional way of life in several ways. On one hand, there was a marked rejection of certain elements of tradition. For example, the advent of refrigeration, commercial freezers, and canned and frozen vegetables made traditional plant gathering and preservation techniques more of a choice than a necessity for survival. Similarly, the diffusion of over-the-counter pharmaceuticals and the implementation of universal health care in Italy in the 1970s put an end to many herbal remedies. On the other hand, some members of the post-war generations revalued past lifeways for their thrift and sustainability, leading to their reclamation.

Agriculturalist Virginio Rota of Vedeseta is among the foremost examples in this category. Finding inspiration in traditional relationships between agriculturalists and the land, he specializes in growing heirloom local fruit and vegetables on his plot at San Michel near Vedeseta. Virginio calls himself a biodynamic farmer, although he rejects Steiner's model of biodynamics in favor of techniques he developed based on traditional Val Taleggio farming practices, including the use of compost, crop rotation, companion plants, and planting 'by the signs'. The latter involves using the phases of the moon as a guide for planting: crops that grow above ground and produce flowers or fruit should be planted on a waxing moon, while those that grow underground (like root crops) or that should not go to seed too quickly (e.g. lettuce and spinach) are planted on a waning moon. He also makes and sells products made from the ingredients he grows, such as jams, biscuits, and pasta made from rye and farro, an einkorn wheat.

9 - A pumpkin grown by Virginio in his biodynamic garden. The plant grows surrounded by dry grass, which allows the soil to be kept moist. Over time the grass decomposes and becomes soil, 2022. (Image courtesy of S. Invernizzi)

Virginio's relationship with trees, plants, and the land preserves something of the traditional reciprocal relationship with the other-than-human world that characterized the pre-capitalist way of life of his grandparents in Val Taleggio. "*La terra mi dà; io devo restituire*," he said ("The land gives to me; I must return"). In his ontology, plants are individuals, just as are all animals, including humans. They each have personhood and are able to communicate. "Some people think I talk to plants, but that's not true," he explained. "The plants talk to me." They speak not through words, but by showing themselves to him—expressing their happiness or unhappiness, problems they may be having, and their general demeanor. As an example, he led me to a 100-year-old apple tree: "What is this tree saying?" he asked. I observed the tree; its leaves were yellowing and covered with black spots; many had fallen off the tree, although it was late spring. I remarked that it didn't seem happy. "Right!" he exclaimed. "This tree has a fungus on its leaves because of too much water" (the spring had been especially rainy and cold). He explained how it is essential to observe plants closely each day in order to be able to interpret their messages. Following Virginio around his orchard and garden, there was no doubt from his behavior that he considered each individual plant an independent being with a 'voice' and a type of personhood. Everything on his farm has a purpose and a relationship to every other element. His worldview is an example of the continuation and further development of a relational epistemology involving the land and the other-than-humans upon it that is rooted in the past, but developed with an eye towards a more sustainable future.

10 - Agriculturalist Virginio Rota of Vedeseta photographed during the research conducted in Val Taleggio in 2016 by Sabina Magliocco. (Image courtesy of the author)

The citizens of Prealpine valleys who are inspired by traditional lifeways and nature demonstrate the cultural resilience of local folkways. They are continuing, in a completely contemporary context, skills, TEK, and praxes that have long been part of this valley. Their worldview, while fully informed by the dominant ontological perspectives of naturalism and materialism, nonetheless contains elements of an earlier relational epistemology.

11 - Elena Brebenel. *Whispers of the Forest*, 2016. (Image courtesy of artist, NAHR archive)

TOWARDS THE FUTURE: THE 'NEW ANIMISM', RESILIENCE AND SUSTAINABILITY

The revaluation of aspects, knowledge and practices linked to traditional lifeways and the ontologies associated with them are more easily recuperated in rural settings. But only a few inhabitants of the Val Taleggio are actively engaged in this form of reclamation. Many others have little direct contact with nature, the land, or pastures and grasslands except as scenery from their car windows. They have fully absorbed the dominant Western ontology in which humans are separated from nature; the other-than-humans in their environment are not granted personhood or agency, nor do they enter reciprocal relationships with them. In this, they are much like other residents of the Val Brembana in the province of Bergamo.

However, as in other parts of the developed world, small subgroups in the region are engaged in reclaiming practices that re-enchant the world and bring about a more sustainable way of life (Taylor 2010; Gibson 2010). For example, modern Pagans living in Bergamo and the broader Val Brembana area belong to a group of new religious movements inspired by the pre-Christian religions of Europe and the Near East, as well as by Indigenous religions (Hutton 1999, 2009; Magliocco 2004). By adapting elements of ancient practice to today's needs, they hope to reconnect to nature, the sacred, and community. Neopagans often identify as members of 'nature religions', in that they believe the sacred is immanent in nature and the material world, and their practices center around seasonal cycles of birth, growth, maturation, death,

12 - Giuseppe Mongiello. *When the Hours Are the Winds in Mantica exercises*, 2022. (Image courtesy of artist)

and rebirth. Since the majority of members are urban, however, most did not grow up practicing reciprocal relationships with the other-than-human, except perhaps with pets. The problem then becomes how to establish such relationships in the absence of any TEK or customary praxis.

Modern Pagans use a variety of imaginative techniques to develop relationships with other-than-human beings in nature and the woods in order to re-enchant the world. Two important techniques are guided meditation and establishing reciprocal relationships with other-than-humans. Guided meditation uses vivid imagery to tell a story in which the practitioner can imagine themselves as a character. Anthropologist T.M. Luhrmann has argued that the use of this technique helps make spiritual entities real for practitioners (Luhrmann 2020). Another technique involves treating other-than-human entities, such as spirits of place, trees, and springs, as though they were members of human society by leaving offerings for them and speaking to them, much as we would to other human persons. This enhances the feeling that spiritual entities are persons with agency and consciousness; it makes them feel real, and enmeshes humans in social relationships with them that entail respect and reciprocity.

13 - Giuseppe Mongiello. *The Straight Furrow In Mantica* exercises, 2022. (Image courtesy of artist, photograph by M. Ranieri)

In participating in activities that enhance the reality of other-than-human entities, modern Pagans may develop ontologies that are more animistic than naturalistic. These ontologies de-center humans, putting them on par with the other creatures, both material and spiritual, who inhabit the planet. While this should lead Pagans to adopt lifestyles that are more sustainable and resilient in the face of environmental crisis, the results are mixed. Modern Pagans say they care more about sustainability, nature, and the land than members of mainstream religions, but in their environmental behaviors, they do not differ substantially from other members of urban society (Kirner 2015). The reasons for this are complex. Neopagans often lack TEK and exposure to practices rooted in a more sustainable way of life. Furthermore, being embedded in a system of global capitalism makes sustainable living difficult to achieve; urban and suburban Pagans remain dependent on fossil fuels and global markets for most of their necessities.

14 - Petra Regent. *Print Mountain Grasses*, 2022. (Image courtesy of artist)

CONCLUSION

We have seen that the inhabitants of Prealpine inner areas of Italy historically cultivated sustainable relationships with their environment because, as anthropologist E.N. Anderson has argued, they combined a deep knowledge of their environment with principles of sustainability deeply encoded in their cosmology (Anderson 1996). Anthropologists studying relationships between ontologies, epistemology and sustainability have criticized Western divisions between nature and culture or humanity, showing that this ontology developed in the West due to a particular constellation of factors not shared by other cultural traditions. These include Aristotelian epistemology, the Christian doctrines of body/soul distinction and the dominion of humans over the natural world, and the development of the scientific method as a way of understanding the world. The resulting ontology, which Descola calls 'naturalism', posits a shared exteriority between humans and other living beings (e.g. being carbon-based life forms; common DNA; shared evolutionary origins), but a very different interiority (i.e., animals and plants are not believed to have souls, agency, or personhood). This epistemology developed hand in hand with a capitalist system that has led to the exploitation of nature for human benefit and an environmental crisis that threatens the planet.

In contrast, anthropologists have documented numerous cultures with relational epistemologies in which humans and other-than-humans are believed to share interiority, and thus to have souls, agency, and personhood, but to differ in exterior form, or in some cases to *appear* to humans as 'other' while actually *being* people to themselves (Descola 2005; Viveiros de Castro 1999). They call these kinds of epistemologies "animistic" (Bird-David 1999; Viveiros de Castro 1999; Descola 2005). Cultures with animistic epistemologies generally have reciprocal relationships with other-than-human beings; they view themselves in relation to these Others and enter into exchanges with them as they would with neighbors and kin. Because of these relationships to the natural world, they generally adhere to more sustainable practices (Bird-David 1999).

Epistemologies and ontologies of human/nature relationships should not be thought of as mutually exclusive, however. In the pre-World War II period, there was a hierarchy of epistemologies in Prealpine Italy, some which could be characterized as preserving a level of animism, others based on religious and scientific views rooted in Western naturalism (Descola 2005). These views coexisted comfortably. The animistic perspective was evident in TEK and reciprocal relationships to the land, nature, and the woods. While not granted personhood as was typical of Indigenous cultures, animals, trees, and plants were seen as living beings that were part of a holistic system; resources were limited, and in order to ensure that they continued to provide what people needed, it was necessary not to exhaust them and to give back to the land. As my interlocutors explained, when gathering wild plants, people took care not to pull out the roots (unless the roots themselves were the desired plant part), and in any case not to decimate an entire stand of a species of plant in order to leave enough for it to regenerate the following year. Farming, too, depended on a system of crop rotation, fertilization, and following so as not to exhaust the earth's fertility. To this TEK was added a layer of naturalism imposed by the Catholic ideology, which presumed not only a distinction between body and soul, but also the exclusivity of human souls and the primacy of humans over all of God's creation. Finally, the introduction of compulsory schooling with the unification of Italy in 1866, and its great expansion in the post-war period, imposed a layer of scientism, or popular scientific ideology, which can often lead people to consider the natural world through a reductive lens, rather than as a living, sentient system. Nevertheless, *some* animals with which individuals or families entered into reciprocal relationships were granted a kind of "flexible personhood" (Shir-Vertesh 2012) that could be revoked if it became inconvenient to view the animal as a person (e.g. if it needed to be slaughtered for food). Herders, in any case, were intimately familiar with their cattle and dogs; in herding cultures, these animals are often seen as persons with individuality and agency (Bulleitt 2005).

The introduction of global capitalism, which expanded into the valley in the 1970s and 1980s, brought about a decline in the subsistence economy, a depopulation of the area, and a loss of TEK rooted in the traditional way of life. The prevailing ontology also changed over fully to the naturalist model of the dominant culture. This model is more tied to patterns of environmental exploitation because it separates humans from other-than-humans, denies the sentience or agency of other-than-human subjects, and is not relational in its structure. In other words, the other-than-human is not understood from a point of view of its relationship to humans, except as a source of income. Relational ontologies are less likely to exploit the land because conceptually, all things, human and other-than-human, share an interiority and personhood. Nonetheless, there are various efforts on the part of local citizens to reclaim and revive older ontological models, practices, and TEK. These include citizens involved in revaluing and reclaiming traditional ontologies and lifeways, albeit in ways that adapt to the existing economic system, as well as those in the broader area (e.g. Neopagans) who use imaginative techniques to create reciprocal relationships with the other-than-human world, embodying a "new animism" (Harvey 2005).

It would be helpful in the future to create exchanges between these two groups so the lessons of the woods and pastures, preserved as part of the Val Taleggio's cultural heritage, can be transmitted to groups eager to learn more sustainable lifestyles. Modern Pagans, most of whom are urban, would benefit from direct exposure to rural lifeways and worldviews, including TEK, to better ground their often-romantic view of the land. Residents of Prealpine valleys, in turn, especially children growing up with little exposure to traditional lifeways, might benefit from learning some of the embodied imaginative techniques practiced by Neopagans to allow them to experience relationships with grasses, soils, trees, rivers, animals, and other-than-human entities inhabiting their environment. Combining praxis-based knowledge and imaginative techniques might yield the best hope for a more sustainable future in the face of global environmental crisis.

15 - NAH_*Workshop* trip to the SOLUNA laboratory, in Brembrana Valley, that was founded in 1921 by the poet, alchemist, and spagyricist Alexander von Bernus, 2019. (Images courtesy of S. Invernizzi)

WE REJECT ALL
DECLARATIONS
OF INDEPENDENCE
OUR EXISTENCE
IS NOT ONLY
OUR OWN

Voices of the Fellows

Grasses & Pastures

Roxanne Korda, Editor

Individual blades of grass together form a community of grasses and pastures responsive to seasonal changes and cyclical economies. The following collection of voices are by human individuals who congregated together, forming a multi-disciplinary community responsive to the pastures and to each other. Explorations of cycles include that of biodiversity of soils, grasses, bacteria; movements of cattle and people across the Italian plains and mountains; cheese-making and curing to form cheeses; animals grazing in pastures; the rich gastronomy of the region. These are inspired voices from varied disciplines and nationalities, each offering their responses to the grasslands.

The summons of grasses and pastures includes reimagining ecological cycles for the restoration of social and ecological commons: a sharing of space. Interdisciplinary designers, artists, activists, dancers, and poets began to awaken deep rememberings of place through responding to space. As the artwork and artifacts become part of a shifting, multi-dimensional cycle, local elements are intersected with global dynamics leading to the emergence of an insight and elucidation which disrupts the quotidian, generating outcomes with each loop of the seasonal cycle.

The following reflections and provocations are a commoning of shared experiences. The archiving of memories in the body, through text, and in images, are a regenerative offering.

1 - Siobhan Rigg. *Reject Independence*, 2019. (Image courtesy of artist)

ON THE COMMONS

Artists have the capacity to observe things, situations, and relationships that don't quite exist or to recognize what isn't widely seen. Given that the capture of categorization and imagination is a continuous part of enclosure, artists offer some potential to imagine alternative economic relationships and forms, in both liveable and fictional ways. The enclosure of commonly held resources, knowledge, and information is an ongoing process. To a certain extent, changes in the conditions of the world also open up potential new areas for commoning. It seems unlikely that a return to forms past is possible or even entirely desirable. One of the things I valued in Sottochiesa [NAH*R*] was learning about the commoning practices in the high pastures and the practices for determining who will graze their animals where each season. For me, that little glimpse of commoning practice in the valley's yearly cycle of the cheese economy was one of the jumping off points for the project that I worked on while there—imagining the grasses and pastures declaring their sovereignty over themselves and establishing initial governance procedures.
Siobhan Rigg, Artist

3 - Jeffery M. Hancock with Northwestern University students Rowan McCloskey, Sophie Teitler, and Ariana Ferguson. *Dancing with Mountain Grasses*, 2023. (Images courtesy of J.M. Hancock)

2 - Cows in the high pastures of Bergamo Orobie Prealps. (Image courtesy of I. Mazzoleni)

Togetherness is important. Shared common spaces are where that togetherness can occur. Artists are uniquely positioned to herald in new spaces, as they are uniquely emboldened to say "I imagine this is here, and so here it is. Please come in." This freedom of imagination and generosity of spirit allow artists to change culture by doing their artistic work. We dance with the grasses.
Katie De Bari, Dancer

To keep up the quality of life of all species, shared public spaces or 'commons' need strong conceptual ideas and attention: they must invite the deep experience of place. The contribution of an artist is to show how this can be integrated into daily life. Public spaces need to move along with all living species around them, creating a sense of co-citizenship, thus building living communities that are resilient and invite participation in, and communal responsibility for, the quality of healthy life for all, physically, mentally, spiritually. In my opinion, the combination of natural grasses and an intermediate level of grazing results in the highest species diversity. For the north Italian Alps region this way of maintaining the landscape will contribute to enlarge biodiversity and reduce climate change.
Floris van der Kleij, Architect

equilibria; indeed, in many ways, language is what allows us to understand ourselves as in common at all. It is in language that we reflect on our social collectivity as a common dialectic moving toward the good of all. It is also in language that we can use this dialectical reflection of sociality to adjudicate, assess, and improve how we relate to (and create, and transform) those practices and beliefs that strengthen our sense of shared public space (and time). Poetry, in particular, helps us understand that language can improve our participation in the social and ecological commons, though it is not always easy. So, in this way, poetry does not demand any 'return to' a shared commons; rather, poetry takes the commons as given. Language subordinated to use, in the sense of language being used strictly for communication, can cause us

4 - The elements of the circular economy of the Taleggio Valley. *Summer pastures near Piani d'Artavaggio*, 2019. (Image courtesy of R. Mead)

Given that the poet works in language, and language cannot be owned, I think most poets have an intuitive understanding of language as a radical commons, a field that none of us possesses, and yet collectively till and care for, not unlike the ecological situation of circulating transhumance in the Taleggio Valley, where grass, cow, cheesemaker, and natural resources produce intuitive emergent to forget that language's beauty is tied to its fundamental identity as a commons. Poetry can unsettle that forgetting, because it is a chaotic process of common production and reproduction, that is to say, it is a *remembering*.
Brent Cox, Poet

5 - Oh-Kyung Jang. *From Val Taleggio*, 2022. (Image courtesy of artist))

ON ECOLOGICAL CYCLES

Rituals are an important way to understand the deep connection between our existence and relation with planet Earth. By building unexpected and surprising thresholds—light to dark, closed to open horizon, from street to garden, from stone to wood, from square to field, from motorway to a field of lavender, a user becomes aware of passing from their everyday world into an experience that is deeply different, thus intensifying their experience and understanding of the world, human and non-human.
Floris van der Kleij, Architect

I am interested in how it might be possible to imaginatively shift (even temporarily) the categories the natural world has been assigned to within western life worlds—where the natural world has historically fallen into categories of use or protection, extraction or veneration. All of this comes out sideways in the grasses project in part because I took it very seriously to imagine the world from the perspective of the grasses and pastures of the Valley. If grasses and pastures exist in formation with the animals that eat them; if they have a relation to one another across altitudes based on the factors that make their existence possible; if they want to deliberate on the challenges before them—how would they go about the pragmatic steps to make that happen? How can they establish a circular system of communication? Where could they access power? How could they participate in the larger administrative governance of the area they occupy? The result is farcical, [I felt left wondering what that result is (outcome)] but the question is serious!
Siobhan Rigg, Artist

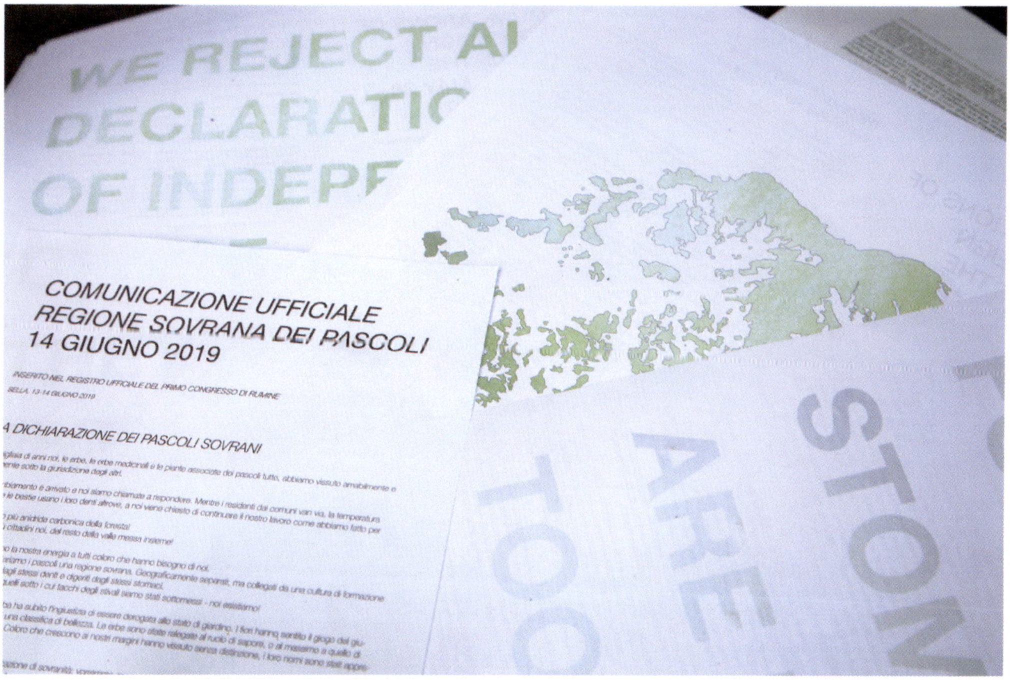

6 - Siobhan Rigg. *Handouts of Declaration of Independence of Grasses*, 2019. (Image courtesy of artist)

Beyond the top of the world

Walking from Sottochiesa to Piani d'Artavaggio

Ten kilometres uphill, tumbledown baite like cairns
doors rotted but those narrow, paddle-shaped feet
still keeping the cows out
the valley shining in surreal green
                                                  far below
ranges stacked back against the horizon.

The people left here are mostly old, still working.
I follow two old men with lean calves
barely keeping pace
on this sun-soaked Thursday.

I stop to catch my breath, inhaling freshgrass-sweetscent.
The cows will go even higher in the last month,
                                                  eat this down to shoots,
                                                  milk rich with sun and air.
I can see all the way beyond the green peaks
                                                  to the glaciers
                                                  to the wolf 's return.

7 - Rachael Mead. *Beyond the Top of the World*, 2019. (Poem courtesy of artist)

8 - Katie De Bari. *Being One*, 2019. (Image courtesy of S. Rigg)

ON REFLECTION

To me, dance is sacred because of the psycho-spiritual and somatic benefits of the act. The earth is required for this—feeling the soil under your toes, having a place to make the shapes and feel the releases that dance offers. Dance can move the earth, draw lines in the soil, bury roots, move the cart, throw the hay... it can nurture ecology as well as nurture our human selves.

I am interested in fostering more of a sacred relationship with the environment through my creative practice by understanding more about the rituals involved in regenerative agriculture and other sustainable systems. How can dance be an instrument of these practices, while also remaining an instrument of human healing and understanding?

...All I know is, I find it much easier, and more enjoyable, to dance in nature than to dance in a vacuum. I can really breathe and be there. There are endless options for movement. I think this is because it's a *communal* space: where the wind can move and I can release voice and insects can buzz by and sun can shine in my eyes and dirt can get in my pores. When I dance outside, I can be a part of the systems I wish to see healthy and thriving. I might even be able to directly support them by dancing within them.
Katie De Bari, Dancer

One of my projects was creating collective kinetic writing exercises to stimulate embodied thinking by engaging with the geography of the residency as a fallow ground for entwining the conceptual and the physical. For example, I created the first exercise after I took a trip to a bocce ball court overlooking a transcendent view of the valley. The exercise consisted of rolling bocce balls back and forth to one another in a variety of ways, while meditating on the actual energy transfer taking place between us. Then we'd write about this experience at certain intervals. One thing I'm thinking of now is how this exercise required not just this local activity on the court with

pasture

from the Late Latin and Old French

past-
meaning grazed

-ure
a suffix forming abstract nouns of action or process
a suffix of feminine nouns denoting employment or result
a suffix forming nouns denoting a collective

together, as a noun:
land covered with vegetation
suitable for grazing

a transitional space, to be passed through

meaning, grazed

9 - Poem written for the Ecopoetry Workshop. Courtlin Byrd. *Pasture*, 2019. (Poem courtesy of artist)

the beautiful view, but that it also meant I had to walk up the mountain, talk with the café owners, rent bocce balls, and engage in this social/community activity, while onlookers wondered just what it was we were doing rolling balls around like kindergarteners. So one of the purposes of these exercises was to help aid us in experiencing poetics as a simultaneously local and expansive, interactive, diffuse socio-physical process without beginning or end. With us, Val Taleggio became not just the *site* of poetry, but an inextricable *part* of the poetic processes underway that we became closer to.
Brent Cox, Poet

A poetic documentary on the process of cheese-making from the pumping of milk generated from the cows' grazing on grass (and the grass's earlier creation by/with sun and soil), to the by-hand practice of stirring curds, forming cheese into blocks, transporting it to the aging facility, and packaging it for sale. Putting this language/definitional work together with the video, a video that was also, in a way, a performance of actual material sociality, helps us see that just the notion of *Grasses, Pastures, and the Regenerative Economy of Cheese* contains within it the sense of the *collective commons*.
Brent Cox, Poet

10 - Courtlin Byrd. *By Hand*, 2019. (Image still courtesy of artist)

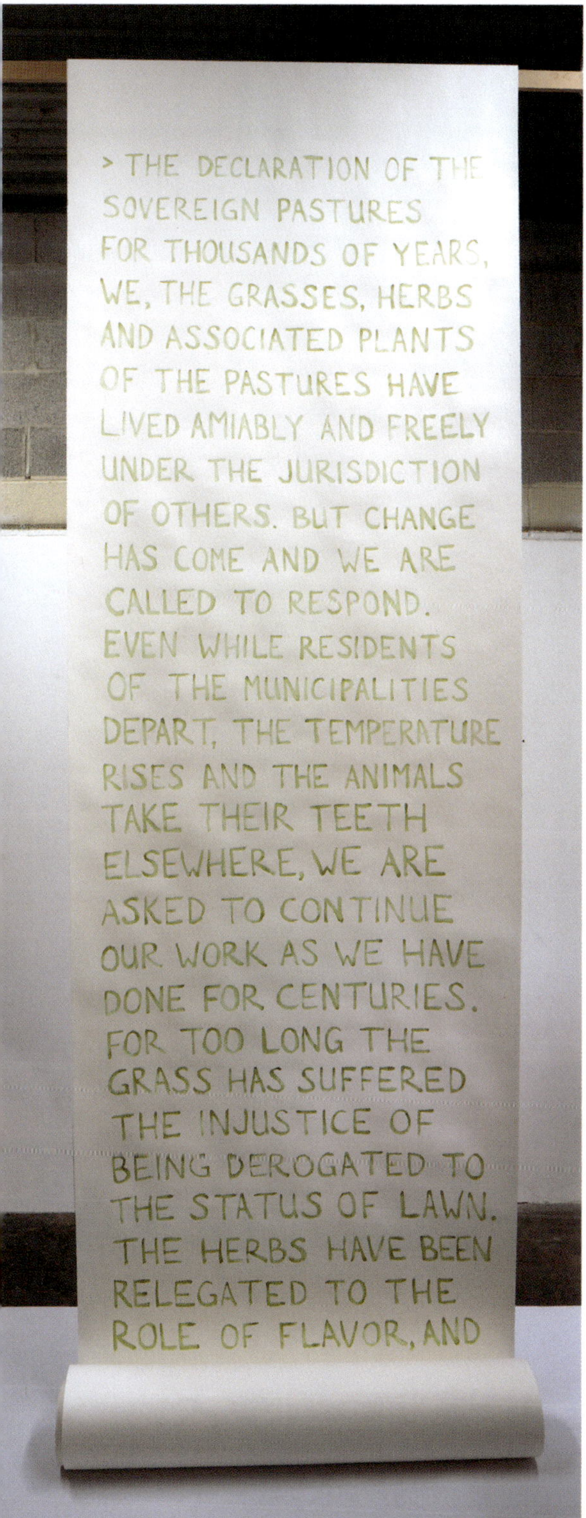

ON CIRCULAR ECONOMY

Considering foraging as an art practice, not just a way of finding food, connects to the way creative practice experiments with how changing and building behaviors alters what you can know individually and communally. Foraging at its most basic is hunting for things that are useful and consumable. As an epistemological strategy though, foraging is a mode of searching.

For me the process of imagining governance for the grass and pastures emerged out of several observations of the sort that you have when you are looking for plants. 1) rumex (dock) is tasty when young for people, but apparently not to cattle—they eat around it, leaving tall robust plants in the middle of a grazed area 2) when the cows move between low, mid, and high pastures they have clear preference for plants they want to start eating first, mixing them with some of the lesser preferred plants from the previous location still in their digestive system.

And two stories about how the pastures continue to exist: 1) the commons aspect to the sharing of the high pastures in the valley 2) the fact that the grazing is a significant part of what keeps the woods back—getting eaten is part of what allows the grasses space to grow. What tends to emerge from those processes in democratic governance are the practices of administration and policy making. One of the central pieces of the grasses and pastures residency was the question of building circular economies. Ultimately, economies rest on input power flows. In this case chlorophyll lies at the root of the valley's cheese economy. And a circular economy demands a logic that keeps each portion of the process in relation with the other parts.
Siobhan Rigg, Artist

11 - Siobhan Rigg. *Long Scroll 2*, 2019. (Image courtesy of artist)

we : cheese
(to keep in touch)
(the microbial relationship)
(client increasingly cold caves)
(density of fat only felt between two fingers look, here, look)
(daily squeezed it is slippery and open to condense something alive)
(love like out of stranger's digestive systems)

in this context what are we
what are "we"
? we are what we
are weeping are we what
wide eyed party constellations
we are cells eat wee wee
are what we eat bacteria
microbiota microbial relation
ship what we eat liebe
geht durch den magen il gu
sto del vivere we eat

12 - Alberto Mazzoleni eating cheese - a 'microbial relation' (Image courtesy NAHR)

Poem written for the Ecopoetry Workshop. A.Monti, *Untitled*, 2019. (Poem courtesy of poet)

Gualtiero Locatelli eating Anais Hazo's cheese on bread, 2019. (Image courtesy of A. Hazo)

13 - Eco-printed plant collection from Taleggio Valley (Bergamo Prealps), Marilyn L. Geary. *The Air Brings Thanks*, 2023. (Image courtesy of artist)

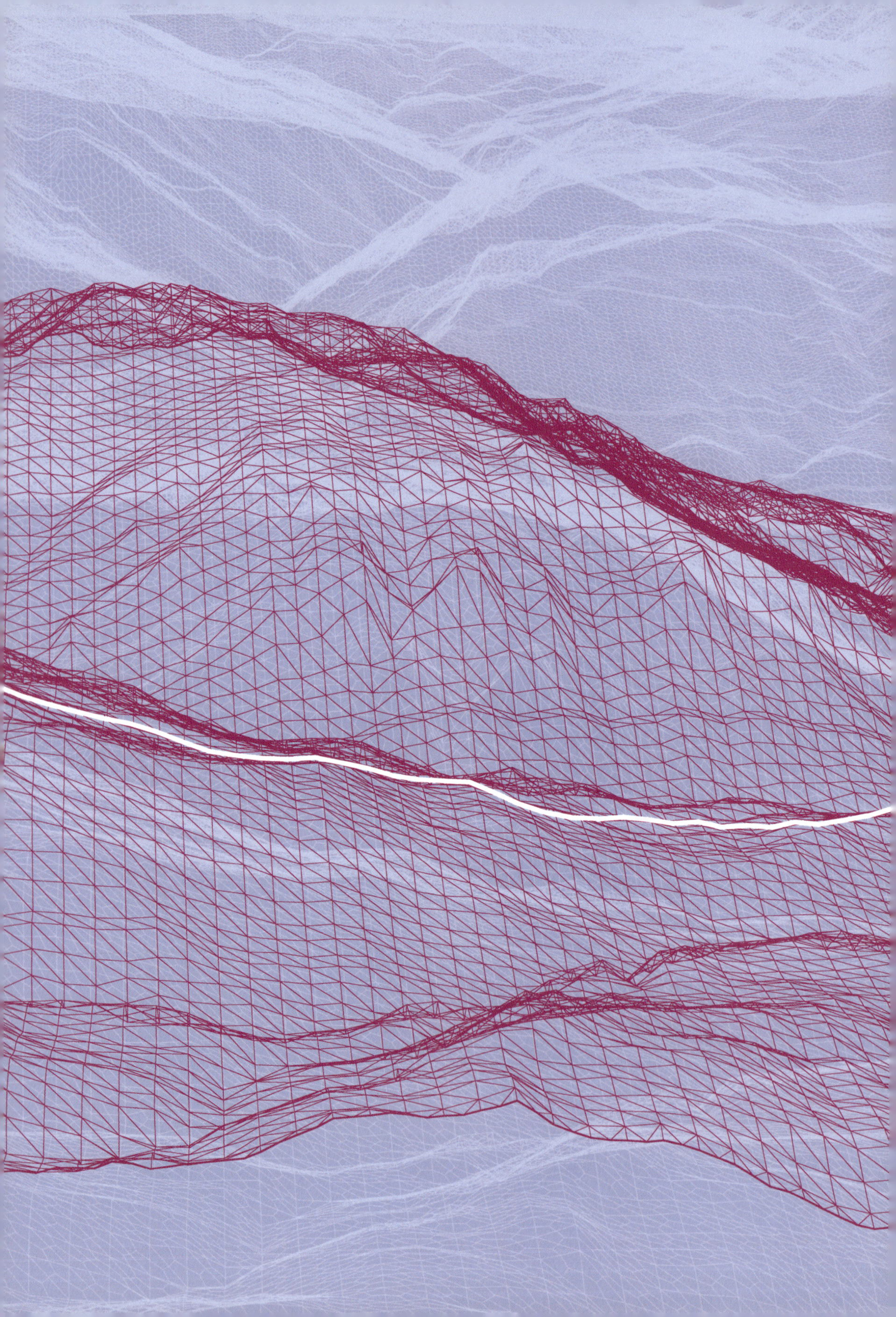

CHAPTER 5

Animals

Interdependence Between Species

Introduction

Gabi Scardi

The subject of animals has always been a filter through which to look at oneself, the dynamics of society, and to question human perception of both otherness and how we relate to one another. In recent decades, so much research has been focused on the human to other-than-human relationship that Harriet Ritvo coined the expression "animal turn" in her text *on the animal turn* (2007). This term has taken on a specific connotation: whereas in the past, within western civilisations, a convinced anthropocentrism led to the representation of the animal in symbolic and moral terms, or as 'the other' to be feared or subjugated. Today, in a post-human framework, we recognise the need to push the boundaries of the human-animal relationship towards a recognition of diversity and greater respect; a mutuality of gaze. The state of the planet itself speaks to the urgency of a new attitude. While sustainability issues may have been considered residual until recently, a shift away from forms of development based on extractivism towards more respectful ways of relating to the Earth and its inhabitants now seems urgent.

As the importance of the issue of coexistence of species, and our pervasive influence on animals, has been infinitely reduced at both the level of the individual subject and the whole, it has also been a driving force in the roots of ecological thinking. It is on this topic, now considered one of the most urgent of our time, that NAH*R* chooses to focus. The themes of ecosystemic relationships between humans and other animals explored in this chapter are in line with an ecological sensibility that has always characterized NAH*R*.

Recently, with the urgency of the COVID-19 pandemic in the background, and through meeting with various key personalities, such as Giovanni Aloi, Stefano Boeri, Lauren Bon, Leonardo Caffo, Armin Linke, Travis Longcore, Ozan Zeybek, and Diana Thater, among others, NAH*R* has explored a vision beyond the immediate health issue to include ethical, social, cultural, environmental, and climate perspectives. The emergent themes of these discussions included the limits of the human being and planet Earth, fragility as a value to be recovered, the contradictions of today's urban order and how it is still possible to imagine alternatives; and again, the incompatibility between certain economic principles and the demands of ecology, alongside the urgency of supporting a concept of justice capable of encompassing all living beings. In an intersectional logic, exploring the topic of animals raises the need to address issues such as biodiversity, coevolution, sustainability, resilience, otherness, consciousness, posthumanism, and climate change. In short, everything we experience in times of crisis requires us to come to terms with our established relationships to other living beings and to search for a more equitable one.

Against the backdrop of the interventions of animalism and situated ecology, an awareness has emerged that the relationship between humans and other animals has always been generative in terms of imagery, thought, and representation. Throughout history, it has not only determined behavior, but has also informed dreams, archetypal projections, cosmologies and mythologies, philosophical and anthropological visions; to the point that more than a few thinkers have theorized that it is precisely the idea of an 'other' animal, seen in opposition to the human 'self', that has enabled humans to see themselves as apart from animals.

The results of collaborative, immersive, and local creative working practices are taken up by the voices of the Fellows in the intervention that concludes this chapter of the book. As for the essays by Sezai Ozan Zeybek, a Turkish human geographer based in Berlin, and Cristina Movalli, an Italian biologist specialized in biodiversity, who has been working in the Val Grande National Park for more than twenty years, they are proof of how many ways there are to narrate the relationship between living things. Their writing examines ways of recovering an awareness of the web of relationships and interdependencies that allow human beings to live.

In *Voices*, the locally produced works of the Fellows, in both the Taleggio and Santa Ynez Valleys, is an observation or, in some cases, active engagement with animals, which has taken on a sensitive form through drawing, painting, installation, textile work, performance, animation or other digital media. It is easy to perceive life in these spaces as a single organic whole, a center of biodiversity. For everyone that is lucky enough to experience such environments, the omnipresence of animals and their clear manifestation, alongside their closeness induced by the small scale, is a personal invitation to renew a fundamental, vital relationship. The common horizon between human and animal and how the animals themselves, from mammals to birds, reptiles, insects and invertebrates, have influenced the environment itself and the way we experience it, becomes apparent. On the other hand, the historical bond between human and animal is still very much alive and here clearly visible, albeit redefined: animals have not disappeared from everyday life as has gradually happened in the cities - except for the most common pets. Human-animal interactions are based on a substratum of knowledge, skills, and practices that have developed over centuries and are still deeply rooted in rural communities.

In the face of a completely normalized human-animal relationship, the free gaze of the creative mind can question and revive perceptions, and generate important, empathetic, and urgent reflections. Among the themes explored throughout this chapter are the vitality of wild and domestic animals, the capturing of the specific subjectivity of individual animals, the possibility of a less utilitarian view of the world and vice versa, the observation of the rural environment and, in particular, certain types of vernacular architectural structures once used for animal-related activities, and many others.

Cristina Movalli presents the critical role parks have through a special case study of Val Grande from which it is clear that only by recognising the paradoxes of human settlement can a deconstruction of the current anthropocentric perspective be initiated. Despite initial impressions, characteristics of rural environments can imply more knowledge but not necessarily more respect. The system of socially organized relationships and the mechanisms of production and reproduction, often regulated by humans, remind us at every turn that ecology is not the same for everyone. In this sense, the park's role is not only to defend a specific ecosystem, but also to promote a new idea of coexistence between man and other living beings. As observed by Movalli, the "aim is that the comprehension of the other as part of a system, rather than something to be controlled by humans, will evolve to allow all species to thrive, and balance between species to be reestablished". In this way, the author emphasizes the close connection between the local and the global. Finally, Movalli invites citizens to adopt a critical and vigilant attitude towards the policy-making entities.

Sezai Ozan Zeybek, on the other hand, goes so far as to question the idea of a completely distinct human individual, circumscribed within himself, endowed with his own unique cosmology and imagination. Drawing on the theories of Bruno Latour and Donna J. Haraway, among others, Zeybek argues that, contrary to what philosophy, psychoanalysis, and legal theory have traditionally maintained, the subject is actually a collection of different entities: a collective. And he proposes to rethink it as ephemeral, dispersed; one agent among others on Earth. Overcoming dichotomous and hierarchical approaches and recognising that the human being is not the only agent of the transformation of reality does not mean, however, shirking the awareness that the transformative power of human beings remains unequal; this entails a strong responsibility.

Focusing on how we, as humans, belong to nature and relate to other species, on how we influence the life of most, if not all, animals—the domesticated, the captive, the wild or the free ones—opens up a vast field of reflection. Such reflections help us collapse the many reductive ways we have categorized and dissected the world, especially in relation to other living beings.

I Exist in Us

Sezai Ozan Zeybek

Radical agency is not about the sharing of boundaries. It is about deborderization.

—*Achille Mbembe*, 'Thoughts on the Planetary: An Interview with Achille Mbembe'

Imagine if we immersed our bodies into a larger network, a more-than-human collective. Perhaps we could open up other possibilities, different cosmologies and new political horizons which would steer us away from the notion of the self-contained individual. This would not just be a postmodern endeavor, which insists on symbolic fragmentations of individuals, but a concrete challenge to politics and judiciary constructs at our disposal, where personhood and agency are conceived as residing in singular human bodies and that all the other beings are devoid of these. What would be the potential of aligning with composite entities that continually enmesh each other and locating agency within a larger, more-than-human web?

I can't help but think: does this body belong to me? There are more bacteria and virus-DNA residing in my body than my 'human' DNA. 10% of my dry weight is made up of bacteria. My immune system, my digestive organs, and many of my bodily functions are entrusted to them. Is this body truly mine?

A multitude of organisms meet, compete, multiply, thrive or die within the space which I presume is my own. They keep me alive (whomever that 'me' refers to) or they may kill me. We are 'thrown-together', as Doreen Massey (2008) once eloquently said. We bring forth a unique, temporary constellation. And not necessarily a coherent one. We move at different speeds, and in different directions, from animal bodies to humans and back. Some survive after I die, and some die so I can survive. I try to set boundaries but in vain. They fly through the holes, crawl beneath the doors, and pass through my skin.

I have ants in my kitchen. Is this house mine?

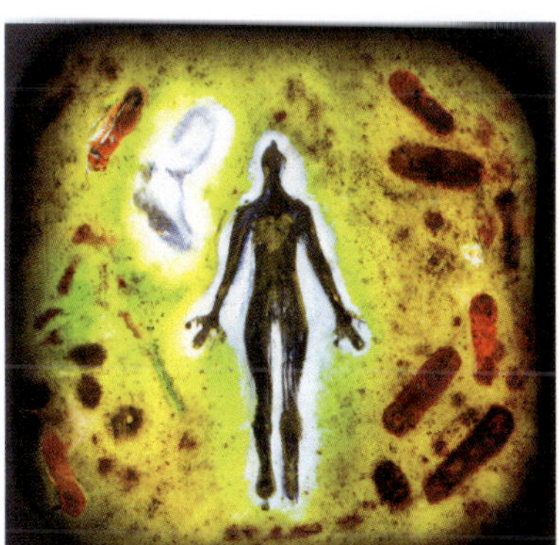

Etymologically, the notion of 'in-dividual' derives from 'indivisible' or 'undivided' and streams down from a certain cosmological postulation. It adheres to the idea of an impermeable, non-divisible being, which is separated from all the other objects, animals, or plants, yet, constitutes a unity in itself. The individual can enter into transactions with others, but it remains categorically distinct from them. In short, it connotes a bundle of integrated personality traits.

1 - Sezai O. Zeybek, created using AI technology: Dall-e-2. *Inside-Outside*, 2023. (Image courtesy of author)

But I know I am more than one. I know it in my guts, on my skin. I am in flux. My dead cells are growing out of my body. I cut them off. They grow back. My existence withers away bit by bit. But I keep breathing in life—not just oxygen, but life forms in their multitude. I take them in. I am a life-world for many. My existence transgresses the judiciary, the Freudian, or the Descartesian self. It overruns my thoughts, intentions and the boundaries of my skin.

No, I do not reside in a collective. We are a collective.

This text will now revisit our notion of nature as a foundational concept, and show its intrinsic conflicts and incommensurability to highlight the dichotomous distinctions it makes between us and them, nature and culture. Then through examining several philosophical accounts that distinguish humans from all 'natural' beings, and particularly from our closest kin, the animals, it will ask how we have ended up with a social world made up of humans only, where others are recessed to 'nature' with no stakes in our 'cultures'?

But we must go further than the critique of the Cartesian divide. Merely looking at social constructs and philosophical accounts pulls the story back to human-only accounts, anthropocentrism and often to Western-centrism. Once again, human imagination and human cosmology become the central causes of existing hierarchies. Again the self-contained individual who *does* the imagining is kept intact.

Is there an alternative vision that would push humans from the center a little to the side and make room for others? Can we position humans differently? Not as unified and coherent but as ephemeral and dispersed. Not as the only agents on earth but as one among many.

We will return to this point at the end.

NATURE/CULTURE

Every time we use the word nature (or natural for that matter), we refer to one of the three distinct sets of phenomena (Descola 2013). In the first instance, we refer to 'unspoiled' corners of the world, untouched by 'civilized' humans, such as the heart of the Amazon forests or the bottom of the oceans. When someone, for example, talks about the 'degeneration of nature', this is what s/he/x implies: an unspoiled nature existing before or outside of human civilization, and where human impact is largely seen to its detriment.

In this schemata, some 'primitive' humans can be relegated to nature, too, who, supposedly, live either harmoniously in the state of nature or lead an uncivilized and barbaric life. The contrasting depictions are derived from the same colonial trope, which was split into two variants from the very beginning: a celebration of civilization on the one hand, which embraced a mission of taming the brutal savage, and regret due to the loss of connection to 'nature', of which the contradistinction was the 'noble' savage.

The repertoire of our emotions and the figures we use to further our claims have not changed much. In the face of the climate crisis, we may be stuck mostly with regret though.

The second use of nature refers to the essence of beings. When someone says that "women are emotional, men are aggressive and that it is in their nature", they make claims in this second order. They fix certain behavioral patterns as natural, and take them as preordained and almost inevitable. In animals, this is called instinct. As if animals can only act from within, in a pre-coded manner.

Yet, time and again, non-human species transgress the nature/culture divide. They change their patterns, they modify their alleged essence. Psychologist Gay Bradshaw, who studied non-human traumas and attachment problems in elephants, observed that young elephants, who have lost their herds at the hands of humans, start behaving like "traumatized adolescents" because they do not receive the necessary in-group 'nurture'. They do not get to learn how to restrain themselves and how to bond with their fellows. As an outcome, they abuse older female elephants, harass rhinos, kill other elephants, and may even harm themselves (Willett 2014). These elephants, like the complex mammals they are, do adapt to new settings, for better or for worse. Thus, natural, eternal, purely biological elephant-ness seems more like a human projection.

Lastly, the third use of 'nature' functions rather as a meta-signifier. Storms, gravity, lions, minerals, earthquakes, and many other dissimilar phenomena and beings are lumped together as 'natural' (as opposed to cultural) and placed on the same ontological plane. They all become the subject matter of 'natural sciences'.

2 - Stefano Torriani. *Gazing Back*, 2017. (Illustration courtesy of artist)

This is indeed an old truism, stretching back to ancient religions. Quoting from *Genesis 1:26*: "And God said, 'Let us make man in our image, after our likeness, and let them have dominion over the fish of the sea, and over the fowl of the air, and the cattle, and all the earth, and over every creeping thing that creepeth upon the earth".

One can see two significant partitions here A) between God and human who is, after all, created in tho imago of God and B) between humans and all the other creatures, which are created for servitude to humans. They make up the 'natural' realm. Even today, such divisions are very well integrated into Western philosophy and scientific conventions. For example, even though the genetic structure of chimpanzees is much closer to humans than to gorillas, they are often grouped with gorillas and as distinct from humans.

A chimpanzee and a human share 96% of their gene structure. With cats, it is 90%. 61% with fruit flies...

Who is drawing the boundaries of kinship and on what grounds? We are a particle in the flow of life. On this small planet, 95% of the total biomass of Earth is comprised of six building blocks: carbon, oxygen, hydrogen, nitrogen, phosphorus and sulfur. This is it! We cannot fully grasp it, but we are made of the same fabric, which was being recycled and transmitted from one generation to another, from worms to dinosaurs, from trees to humans. We are droplets in a great river. We resemble each other a great deal. We grow, we process energy and we break down. Alas, we drown in the difference!

3 - *Kindreds*, 2019. (Image courtesy of I.Mazzoleni, NAHR Archive)

A SHORTCUT TO METAPHYSICS: THE PHILOSOPHICAL ROOTS

The flipside of carving out nature as distinct from the (civilized) human realm, is to abstract humans from the rest, from animals, plants, and, at times, from the animal existence of humans themselves, i.e. from their own bodies. These are moves operating in tandem, in which spirit, reason, and complex language have been warranted as the three main distinctive identifiers of humans. A recurrent dividing line emerges in various forms.

The first dividing line, the spirit or soul, was claimed to be laid with Socrates' proposition saying that humans have immortal souls, which are innately different from the body and transcendental to physical existence. Accordingly, the soul is considered to be superior to the body to the extent that it could surpass its carnal limitations.

The soul is the abstracted essence of humans—devoid of materiality. Pure. No blood, no flesh, no carbon, no oxygen. The soul is supposed to be the real me. Enclosed. Nothing dispersed, nothing traversed.

In a similar vein, reason, too, is juxtaposed against nature in a hierarchical relation, which is then reproduced in other dichotomies as well: men to women, civilized people to savages, human to animal, and so on. By comparisons and contrasts, by rupture and difference, these archetypal dichotomies and value judgments flow from one cluster into another through transferences and through analogies: savages being equivalent to children, children to animals, and animals to all unreason. In the end, dichotomous thinking, it is claimed, constitutes the ontological underpinnings of Western cosmology, which classifies the world in a hierarchical manner.

In this ordering, animals, which are supposedly not granted a soul, language, or reason, and which are taken as 'nature' itself, could only bear a corporeal existence, and this becomes their defining feature. Their utter corporeality deprives animals not only of an afterlife but even of death. Martin Heidegger (1982), for example, claimed that animals could only perish [*verenden*], not die:

> Mortals are they who can experience death as death. Animals cannot do this. But animals cannot speak either. This essential relation between death and language flashes up before us (p. 108).

From Kant to Hegel, from Heidegger to Blanchot, animals are denied the truth, the manifestation of soul, space, time, or even death. They are depicted in a perpetual state of 'unknowing', which could only give automatic responses to their surroundings like Descartes' *automata*. They functioned as the constitutive others and marked the boundaries of humanity and arguably inscribed purpose to human existence. As Maurice Blanchot noted: :

> Death, in the human perspective, is not a given, it must be achieved. It is a task, one which we take up actively, one which becomes the source of our activity and mastery. [Man] makes his death; he makes himself mortal and in this way gives himself the power of a maker and gives to what he makes its meaning and truth (1989, p. 95).

While 'man', the stand-alone entity, could achieve death and elevate 'himself' to higher meanings, the animal belonged to the great unknown and was lost in the void once it 'perished'.

I wonder, who says that humans are at the peak of evolution? Life continues in utterly different forms. Millions of years later, our primitive lives could offer deplorable lessons to others, why not? Many great creatures perished before us. All other Homo species

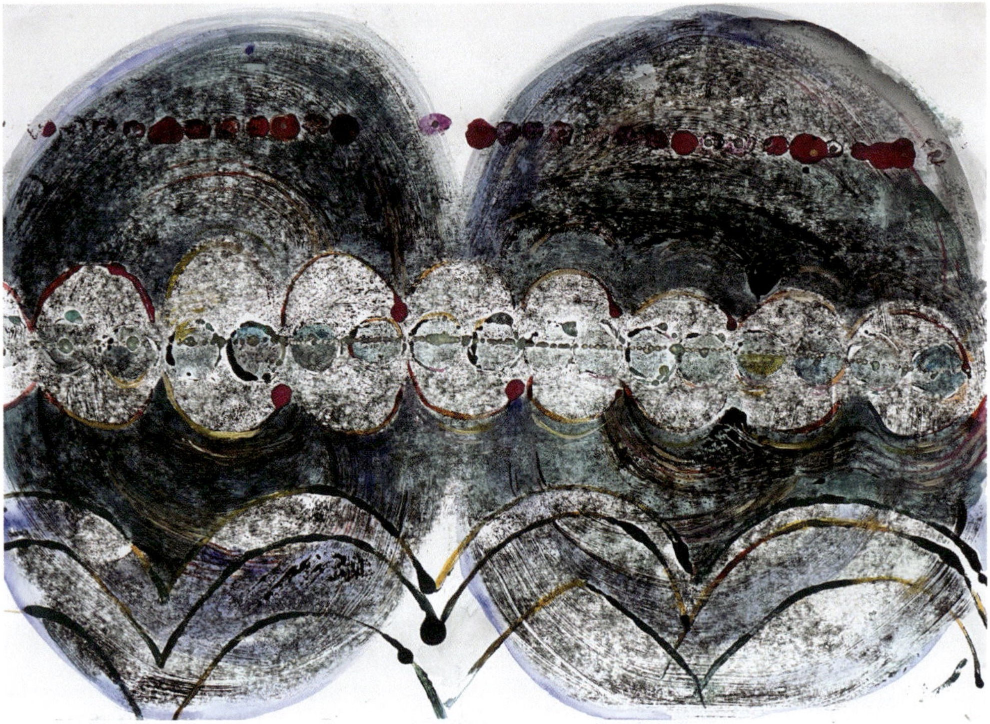

4 - Heartbeats of Living. Sarah Valeri. *About 12 seconds*, 2022. (Image courtesy of artist)

before us went extinct. What guarantees the sapiens will be the exception? Who says we are the final station?

In a similar vein, Hegel (1975, cited in Timofeeva 2018, p. 62-63), although hinting at the existence of some sort of 'inner life' in animals, found it worthless all the same.

> Animal life is throughout restricted and tied down entirely to its specific qualities. The sphere of its existence is narrow and its interests are dominated by the natural needs of nourishment, sex, etc. Its inner life is poor, abstract and worthless . . . The living thing in nature does not reveal its soul for itself, for the thing in nature is just this, that its soul remains purely inward, i.e. does not express itself as this ideal unity.

We can come up with more examples like this from various philosophical schools. The common ground in all these is that non-humans are depicted as deprived of *logos*, language (*phonè semantikè*), mental images and representations *(mimesis)*, and the capacity to reveal the truth *(aletheia)*.

Many activists and scholarly works concerned with violence against non-humans set out with a critique of this Cartesian divide, which, they argue, denigrates non-human animals and reduces them to a disposable physical existence. Consequently, the hard-to-imagine treatment of animals in industrial food production or those who are sacrificed in scientific experiments, or entertainment, are explained by this intrinsically Western ontology.

Yet, the problem with this account is that societies that have harmed animals are not confined to the West. Outside of the Greek-Judeo-Christian tradition, one also finds systematic animal killings, sacrifices and even wiped-out species.

After the first humans reached New Zealand about 800 years ago, the majority of the local megafauna went extinct, along with 60 percent of all bird species. Archaeological findings indicate similar faunal changes in Polynesia, the Caribbean, Madagascar and the Americas. In the latter case, "within 2,000 years of the Sapiens arrival, most of these unique species were gone. According to current estimates, within that short interval, North America lost thirty-four out of its forty-seven genera of large mammals" (Harari 2014, p. 79). The Roman gladiators entertained the public by slashing through wild predator animals. Ottoman Sultans (and other 'Eastern' emperors) rounded up wild/exotic beasts and kept them in cages to fascinate and scare visiting diplomats. Even today, during the Dashain Festival in Nepal, hundreds of thousands of animals are sacrificed to the Goddess Durga. In other words, not everyone in the West came to terms with animal killings, nor do non-Western societies turn out to be placid in their dealings.

Besides, the divergences of European thought are often ignored when the Cartesian divide is posed as the only basis of European thought. To cite one, David Hume (2007) argued that the difference between animals and humans was a matter of degrees, not of kind. Neuroscientists working on animal communication, or philosophers who contemplate transitions between animals and humans would also pose exceptions to so-called Western binaries.

To talk about violence everywhere is not to say that humans have a natural tendency to excessive violence. The distribution and degrees of violence greatly vary among human societies and across time. The manifestation of violence does not follow a uniform pattern or hint at an unchanging human nature. The point raised here is that framing violence, or any type of hierarchy, as a function of the Cartesian divide, necessarily draws back on a selective Eurocentric literature. As such, not only Enlightenment, progress, or democracy, but also excessive violence and abuse of non-humans appear to be rooted in the so-called Western civilization.

This is another version of diffusionism, which James M. Blaut (1993) succinctly criticizes for turning the rest of the World into a mere recipient of what has already emerged in the West. Its impacts may be evaluated positively or negatively, yet in either case, the West appears to be the driving force of history.

Therefore, we would suggest going beyond the metaphysical or culturalist explanations, to steer away from the question of 'why' and focus more on the questions of 'how', 'by which means', and 'by whose participation'. This would, I claim, take us further than solely focusing on the mental categories of humans and open pathways for grasping more-than-human collectives better. The capabilities of a society, the type of networks it builds, the materials it can mobilize, and the bodies it brings into contact involve non-humans (objects, plants, animals, and other hybrid beings) at every step of the way. Understanding the causes of violence merely at the level of mental constructs is like trying to calculate the destructive power of an army by its soldiers' state of mind—without looking at their tanks, missiles, satellites, or logistics.

MORE-THAN-HUMAN COLLECTIVES

Bruno Latour defines "collective" (as different from social) to foreground non-human agents and relations. According to him, longstanding sociological categories such as the nation, family, class, neighborhood etc. are often treated as givens, more or less as stable categories consisting of humans and human relations. The ordinary sociological inquiry starts with limiting the number of actors, often by excluding non-human beings, or technological materials, and how they eventually shape the assemblage.

Instead, Latour argued, the starting point of sociology should be to study case by case how any category comes into being by different forces and is constituted by the participation of different actors. In other words, every inquiry should thoroughly explore who and what participates in the action. This means letting non-human elements in. For Latour (2005), when we act as enquirers, there is a political, scientific, and moral imperative to not make assumptions about what or whom our societies are made of. His formula is to allow the actors to lead us, whatever they may be.

A large chunk of my memory (photos, emails, my diary) is stored on my laptop. My love comes to fruition in the long letters I have written. I need my glasses for my vision. Without those, I am less. As a matter of fact, without those, I am not me. We are always extended to other beings and other objects. The capabilities that I think I have, even simple ones such as reading, mean nothing if there are no books or no words. When I enter the market, I look at the price tags and read the packages. I do not possess permanent economic structures. I perform them by engaging with these labels. As plug-ins are added to the software, tags 'update' me on the spot with new criteria. Economic principles are coded into our surroundings by many such plug-ins because acts are realized in this larger environment, through the assemblage of a composite of beings. In such compositions, sharp distinctions between subject and object, inside and outside, between culture and nature, or human and non-human lose relevance.

I am realized with others. Not just with constitutive symbolic others, as in me vs. you, but with the multitude of beings around me.

'Human' is not the sole agent of such collectives. Usually, we do not have a full grasp of the impacts of the collectives either. Humans are nodes in larger networks.

However, this does not mean that all nodes have the same power or the same effect. Some components of the collective affect the lives of others asymmetrically. Their capacity to organize or modify relations is not the same. Humans have been transforming the earth by using fire for at least 400,000 years, or by agriculture and science. They influence their environment immensely.

5 - Sezai O. Zeybek, created using AI technology: Dall-e-2. *Entanglements*, 2022. (Image courtesy of author)

But this does not ascertain that humanity is at the top of the chain of life either—if we are to assume a peak point, of course. For 3.5 billion years, bacteria have modified everything from stone to the soil, from humans to plants. They transformed the entire face of the earth. They changed the composition of the air we breathe and the content of the water we drink. Some bacteria can kill, or let lives thrive. As a species, they are older, more resilient, and have had more impact on this earth than humans.

Conscious or not. The important point is that they have changed the composition. This is what agency means. They do not merely predictably transmit an outside force, like billiard balls do.

It should also be kept in mind that the transformative powers of human beings are not entirely due to their conscious acts. Indeed, many of humans' long-term impacts on Earth will apparently be a result of unforeseen, unplanned, unconscious activities. In short, consciousness is not a requirement for changing the composition.

6 - Terra Formers. Fulvia Mendini. *La mano di Buddha*, 2012. (Image courtesy of Cemon)

CONCLUSION

What we call collective is formed by the assemblage of unequal components, by the gathering of beings that can exercise asymmetrical powers.

Humans are not unchanging entities. Their capabilities, mental categories, and morphology have changed many times within collectives and still are. Our domesticated companions changed our mobility, our diets, and our technical capabilities. We may soon jump/evolve/morph into other planes of existence through gene modification or new machine interfaces. As Donna J. Haraway tells us, "[i]f we appreciate the foolishness of human exceptionalism then we know that becoming is always becoming with" (2007, p. 244).

7 - Many-headed civilisation. Alessandro Mendini. *Mostro 1*, 2014. (Image courtesy of Archivio A. Mendini)

Although singular acts may seem insignificant at first, their accumulative influence over a longer period could be vast. Every action is performed with the participation of different beings, be it satellites, worms or cotton. Not by individuals, and not only by human individuals. Acts are common acts. Hence, the transformative potential of acts, thus agency, is not limited by human activities only, or by consciousness, logic, and spirit.

These postulations have been put into practice by so many different societies throughout time, where, as in animism, the separation between humans and non-humans was not as strict, where humans were animalized or animals humanized, and where consciousness had not been seen as springing from self-reflexivity of one individual [*cogito, ergo sum*], but through a larger, polyvalent, network of dispersed relations.

We do not have to revert to one of these models. But we can get inspiration from other paths that have been taken and could have been taken.

One last addition. Living as a collective does not imply that I should hold myself back. Even though this text was a call for humility, I still have a lot to say in matters that concern us. Because when it comes to the climate crisis or corrupt governments, we are not on an equal footing with fruit flies. We have more power to modify the collectives, exercise more force, and think ahead, even if they fail at times. We can craft ourselves what to be, but in the knowledge that we are trans-bodily and intersubjective beings.

Because I exist in us, and you in me.

8 - Cornelia Funke. *Magpie*, 2003. (Illustration courtesy of artist)

Our Unique Mountain Homes

Cristina Movalli

I believe...that our national parks, our parks generally, but especially our national parks...are logical centers for both scientific research and education. For many domains of science but especially and critically for biodiversity and conservation of the living part of the environment.

—*E.O. Wilson*, 'Mission of the National Parks and its Relevancy Today'

Life on Earth is sustained by its biodiversity. Selected territories are today regarded as critical locations where such diversity can and must be preserved, through several key actions. Humans must act as ambassadors to the other human and non-human animals making sure to respect and protect these invaluable environments and ecosystems. Protecting life in the mountains means exactly to protect its biodiversity.

1 - Golden eagle. *Aquila chreysaetos, juvenile*, 2020. (Image courtesy of M. Monfrini)

There are three levels of biodiversity. The first is diversity of ecosystems, referring to the different environments in which life is present, such as forests or deserts or mountains. The second is specific diversity (the most commonly used meaning), that is, the complex of species that inhabit a certain region. Thirdly, there is genetic diversity which is the variety of genetic information contained in different individuals of the same species.

The great biodiversity of the Alps is today threatened by human activities on both the local and global scale. This has been recognised in various international declarations (cited) as the Alps are divided across different countries. They have united to establish areas of protection (the parks) and later actions (declarations) based in scientific research, education, and dissemination.

The Alps are home to more than 30,000 animal and 13,000 plant species (The Alpine Convention 1991) which is an extraordinary richness of specific diversity constituting an extremely biodiverse zone. Extreme environments from high altitudes (Mont Blanc is the highest peak at 4810m) are dominated by ice and sub-zero temperatures that host very specialized living beings with remarkable adaptations. Together with the value of its richness, we need today to consider the fragility of these mountain ecosystems.

2 - Capercallie. Stefano Torriani. *Tetrao urogallus*, 2017. (Illustration courtesy of artist)

Human activity such as hunting and poaching, as well as increased pollution and land use change have significantly impacted and threatened the survival of the animals and species in the Alpine region. Alongside these local factors the vulnerable and fragile mountain ecosystems are affected by human-made global changes and subjected to climate change events. For example, the dramatic reduction and melting of glaciers, has caused flooding and drought, changes to river cycles, and the fragmentation of habitats. These have had effects both on the mountain and downstreams causing some species to be displaced to higher ground or even to become extinct. These

3 - Female chamois. *Rupicapra rupicapra, with its young,* 2021. (Image courtesy of S. Crocetti)

concerns and realities are well known now to the Alpine nations, and have been laid out as principles by the *Declaration of the XVI Alpine Conference on the Protection of Mountain Biodiversity and its Promotion at International Level* (2020).

For this reason, the protection of the Alpine environment through the establishment of protected areas is of fundamental importance. As established by The Alpine Convention in 1991, and described in the *Framework Convention: Guiding principles for sustainable life in the Alps* (n.d):

> the objective is to protect, conserve and, where necessary, rehabilitate the natural environment and the countryside, so that ecosystems are able to function, animal and plants species, including their habitats, are preserved, nature's capacity for regeneration and sustained productivity is maintained, and the variety, uniqueness and beauty of nature and the countryside as a whole are preserved on a permanent basis.

Over time, with great effort, the European countries that share this mountainous territory have been able to protect almost 30% of the Alpine region by instituting several national parks and other forms of natural conservation bodies (regional parks, natural reserves, naturalistic oasis). In Italy, the first national park, that of Gran Paradiso, was established in the Alps in 1922, followed in 1935, by the Stelvio National Park. Only after several decades, two other important national parks were created: the Belluno Dolomites, in 1990, and the Val Grande in 1992. The existence of the parks puts into practice the new Law 394/1991 on protected areas as they assume a strategic role in the environmental conservation of species and habitats, and sustainable development policies to assure an ecological balance.

4 - Gran Paradiso National Park, the first national park in the Alps, 2019. (Image courtesy of M. Monfrini)

In the meantime, turning to the wild or natural (although rare and hard to find) we have modified environments with our actions and lifestyles as mentioned above. The habitat alterations have had lasting negative impacts on many species. Not only do the direct threats, such as habitat fragmentation, intensive human activities, and land consumption cause extinctions, but also the abandonment of traditional and less impactful agricultural practices can lead to a loss of biodiversity (e.g. in mountain grasslands). Post-World War II urbanization led to human migratory patterns from the mountains to the plain. Development of these large metropolitan areas left more human-free spaces in the mountains (some of which have been converted to parks) which indirectly facilitated the repopulation of wildlife.

Moreover, parks are recognized as fundamental places in which to integrate the needs of nature with those of humans, enhancing and experimenting with compatible productive and recreational activities, promoting scientific research, and actions of education and dissemination.

SCIENTIFIC RESEARCH

Due to the increase in endangered and extinct species in the mountains, caused by human activity (such as hunting, fishing, poaching, and land use) the park's key role has included fauna reintroduction projects in the case of extinct species, and repopulation projects when the dynamics of the populations do not guarantee their survival. There has also been an effort to try to eradicate threatening and invasive alien species. Individual institutions learned that just as everything in nature is interconnected, similarly policies and interventions become more effective if implemented from a system perspective. Program agreements across institutions and memoranda of understanding, developed between various public and private subjects, aimed at solving common problems by establishing cooperation efforts. This has even occurred across countries for the natural world is not bound by our borders. Indeed, only through research is it possible to know the health of an ecosystem, monitoring changes and, consequently, providing the data necessary to plan actions aimed at their protection.

5 - The Grey Squirrel, *Sciurus carolinensis*, is an example of invasive species in Europe. It was introduced from N. America in 1867 to England and in 1948 to Italy; it is responsible for the decrease and even extinction of the native Red squirrel, Sciurus vulgaris, because it is larger and more robust with a competitive advantage in the search for food, 2022. (Image courtesy of M. Monfrini and S. Crocetti)

6 - Does the wild animal want to be touched, 2021. (Image courtesy of P. Nalin)

ENVIRONMENTAL EDUCATION

Considering from both this ecological and regulatory context we can think of two main kinds of 'nature' that humans can expect to encounter today. One is the bucolic, postcard-like view, with grazing cattle, well-kept restored buildings with flowers on the balconies, and beautiful landscapes all around. The second is severe, something sublime, without a permanent human presence: in one word, just Nature. The second is most obviously the best place to encounter wild animals. Everyone during a hike wishes to observe in some measure a wild animal but what is the best approach to observe it?

One can have a passive, contemplative, and discrete predisposition, taking a careful look from a distance considered safe (which varies from individual to individual). Others might want to enter into a more direct and close relationship with the animal, even trying to touch it. However, as humans we must ask ourselves: does the wild animal want to be touched?

Whether consciously or unconsciously seeking this experience we can still expand our limited toolboxes to aid our encounters with the non-human. In a zoo where we pay for a ticket to see animals, the meeting can be taken for granted. On the other hand, we are visiting prisoners! In the wild, things are different. The discretion is crucial and the encounter is not a certainty.

When responding to the question 'does the animal want to be touched?', we must consider if it would rather have enough space to live freely, without roads cutting its territory, able to find prey and to be preyed upon, without having humans managing their reproduction activities and food availability. When do humans ever really position themselves to look at things from the viewpoint of others? Not simply wanting to make decisions for the others, but rather giving enough space to the others so as to allow evolution to run its course? Even though we mostly view visiting parks as a recreational activity, gaining an environmental education by reminding ourselves of these questions, whilst framing them in the natural and wild environment that has been impacted by human activity, is really what the park setting is all about.

Educational research allows human animals to evolve their understanding of the other, non-human animals. Many humans express phobic reactions to nature and certain animals, while the more biophilic among us crave these interactions and are intrigued and delighted by the wild and natural world. Though it may appear a banal question to some of us, we know naturalists such as David Attenborough, Jane Goodall, and Piero Angela (to cite only some) have dedicated a great part of their lives to it. Rangers and environmental guides in parks do the same.The role of park education programs is to help us understand the ecological role of all living organisms, of 'the environment' as a whole and of its interactions. That is why, even within regions like the densely populated area of northern Italy, where the species are still confined within a generous, but limited territory, we can see the value of such spaces today. Education is the first step to understanding that the 'other', even when different, cute, or dangerous, still needs respect. This experience of nature reconnects us back to our instincts as animals too, and reminds us how we are all interconnected within one whole environment, and together are hosted by planet Earth. The parks allow us to reintroduce, respect, and begin to understand some of the species which we might never consider but which are critical to life as a whole. These park education programs have the power to transform the visitor to an active participant in the conservation of these ecosystems. They can transform from visitor to 'Citizen Scientist.

ENGAGEMENT AND DISSEMINATION

7 - Appreciating nature through illustration. Nora Sweeney, 2021. (Image courtesy of artist)

Today, one witnesses a contradiction in the relationship between *Homo sapiens* and other animals. On the one hand, we humanize our pets by anthropomorphizing and treating them like humans. Whilst on the other hand, we are attracted by the untouched wilderness as a symbol of Nature. However, we also want to manage or eliminate an animal when it annoys, frightens, or interferes with us and our activities. We want to coexist, but only in the way *We*, as humans, can dictate it. This vision was emphasized during the COVID-19 lockdowns when people, locked up at home, rediscovered the need for Nature which led to an overcrowding of natural spaces near cities as soon as it was allowed. An opportunity has now arisen as new groups of people have developed who understand the criticality of 'nature' with an evolved and more respectful consciousness towards it. It is now understood on a broader scale that we are in fact all stakeholders in these rural and wild environments, even if the majority of our lives exist in the urban setting. Some key programs such as 'Citizen Science' seek to both educate the general public in the destructive issues surrounding life in the Alps, as well as engage people in positive actions to collect data, to regenerate the ecosystems and heal the wounds of our actions. This new critical mass of educated and aware voices will also help to push through important changes in legislation by supporting the campaigns launched by local activist organizations.

VAL GRANDE NATIONAL PARK - A WORTHY PRECEDENT

The Val Grande National Park is an emblematic place to understand some of the actions taken to protect the biodiversity of the Alps. The Park is located in Piedmont, in the northwest of the Italian Alps. It is a protected territory nestled between densely populated metropolitan areas such as Milan and Turin, and natural tourist attractions such as Mount Rosa and Lake Maggiore, bordering Switzerland. It was established in 1992 as it was recognized as the last great wilderness area in the whole European Alps.

When focusing on the word 'wilderness' our thoughts generally turn to remote spaces, uncontaminated by the presence of human beings. But Val Grande is different, here the traces of human presence are very evident. Wilderness in Val Grande identifies itself mostly as an abandoned human settlement, where nature, little by little, for over 70 years, has reclaimed the land as its own. In the park there are no roads and no telephone coverage, it is the reign of silence and darkness.

This may feel charming or mysterious and the solitude may feel attractive to biophilic souls, however isolation may be scary and phobias can start to become overwhelming to others. Today, the Park offers invaluable space to many species that can live freely. In this mediation, gentle observation is allowed to humans. It is a place for respectful recreation. For researchers, it is a valuable space for continued support of biodiversity regeneration.

Today, many animal species thrive, provided there are actions put in place, including repopulation or reintroduction projects. When the area was inhabited by humans there were mostly domesticated animals. Nowadays, it is possible to encounter both the celebrated and adored animals such as a female roe deer with its cubs or a golden eagle flying above the canopy, and also those animals that historically or culturally were perceived as dangerous like snakes, mice, bats, caterpillars, bees, wasps, and insects in general.

8 - Example of rewilding. Vegetation covers and hides the remains of the ancient peasant civilization, 2008 (A). Alone in front of majestic and secular beech trees: wonder and contemplation, 2014 (B). (Images courtesy of author)

The creation and dissemination of environmental protection policies, national laws and European directives, along with the consequent management of fauna, including better hunting practices, have also favored the recovery of some animal populations. Today our Alps and Prealps are rich in ungulates such as chamois, roe deer, and wild boar, which constitute potential prey that have contributed to the natural return of a predator such as the wolf.

An emblematic and well known monitoring project is in fact related to the most important predators of the mountains: the wolf and the bear. Ancestral stories, fairy tales, and our own human animal instincts have instilled a sense of fear and phobia in relation to the wolf. In the Alps it went extinct due to hunting around 1920 and during the sixties and seventies started to move from the Apennines, scattering to the North looking for suitable and available territories. Now, since the recolonization of their territories back available and a land use change, wolves are quite well spread across the Alps, especially in the western area. A European project within the LIFE programme, called *Wolfalps EU* is dedicated to "Coordinated actions to improve wolf-human coexistence at the alpine population level" (LIFE Wolfalps EU n.d.). The planned actions include a standard monitoring that can be done with camera traps or snow tracking. If lucky, the researchers can collect samples for genetic analysis in order to study the whole wolf population.

Since 2000, wolves have appeared, dispersed, and moved in search of a territory across the Verbania region. However, finally in 2019, the first partnered pairs were seen, proving reproduction was possible. Even In Val Grande National Park the wolf has now reestablished its presence in 2022. The return of the predator has inevitably caused many conflicts, especially with farmers. In the last two years, the local press have emphasized the farmers' rage. We are trying to find a solution to coexist with the predator, but many farmers remain angry despite the prevention systems and the opportunities for dialogue.

Another predator has recently appeared in the area: the bear. But bears do not come back naturally in the Alps; they were reintroduced in Trentino with specimens coming from Slovenia. Now in the Alps, bears total around 100 (Groff et al. 2022). Thanks to genetic analysis most bears are identified with a specific code. In April 2021, a park ranger found footprints in the snow, proof that the bear was there, passing by. But the bear with code M29 could be anywhere now! The problem with bears is that sometimes they can be too confident when they get too close to inhabited areas looking for food in rubbish bins, or beehives. The bear can be more dangerous than a wolf just for its size and even hikers are worried for their safety.

9 - A crayfish in a stream of Val Grande National Park, 2021. Researcher conducting a microchip marking operation to help find the crayfish again, 2021. (Images courtesy of M. Tessaro and author)

Below we will further explore some more of the current and past programs which have been successfully implemented in the Val Grande to aid in the goal of nature conservation. These will cover examples of extended research and dissemination to the local communities and visitors offering insight into a successful and ongoing implementation of policies and conventions. Learning about these various interventions and studies provides an inspiration from which to apply our own actions to environments around the world, both in urban and rural districts, to better understand and care for the ecosystems we inhabit.

1. The Reintroduction of the White-Clawed Crayfish (*Austropotamobius pallipes*)

Thanks to the LIFE Nature program, the Park was a partner of the *IdroLIFE Project* which also financed an activity survey and subsequent restocking program of the white-clawed crayfish (*Austropotamobius pallipes*). Its area in recent years has greatly shrunk because of pollution, competition with alien species, and climate changes. A preliminary study obtained an unfortunately unsatisfactory picture of the distribution and health status of crayfish due to the high fragmentation of the populations and their inability to maintain themselves in the long term. Even the serious water crisis, which is a direct impact of global human activity and climate change, has affected the entire catchment area of the Park. This has been evident during the past summer seasons of 2022, and has most likely contributed to aggravating the state of vulnerability of the species (IdroLIFE n.d.).

Management strategies aimed at protecting the current populations present which are at risk of extinction, were therefore implemented after identifying the best suitable sites for a restocking action. Many of the released individuals, taken from areas adjacent to the Val Grande area, were marked with a microchip so as to be able to monitor the effectiveness of the intervention over time.

2. Monitoring the *Phengaris alcon*

Val Grande National Park has been engaged for about three years in research dedicated to pollinating insects whose survival is also closely linked to our very existence. Particularly interesting among these is the investigation conducted on *Phengaris alcon*, a butterfly present in a very localized way in a few places in the Piedmont area and threatened by both the change of its habitat due to land use changes, and climate change. When animal species are generalist and ubiquitous, they are more able to adapt to the changes taking place, but this species of butterfly is very specialized and therefore more vulnerable.In fact, its survival depends on some fundamental elements: the presence of its nurse plant, the *Gentiana pneumonanthe*—which grows only in wet meadows and heaths from the plains up to 900 meters above sea level—and the presence of ants of the genus *Myrmica*. The adults lay their eggs exclusively on the aforementioned species of gentian and only on the buds of the plant. The larvae that hatch then move away from the flower to settle like parasites in the anthill that

10 - A butterfly specimen marked with a specific number allowing for tracking and unique recognition on recapturing, 2022. (Image courtesy of S. Nelli)

will host and feed them throughout the winter. The following summer, the adult butterflies will take flight to look for more gentians to lay other eggs and ensure the continuity of the species, provided that the environmental and climatic conditions have allowed the presence of the gentians and the ants to parasitize.

A survey was conducted by a team from the University of Turin in a meadow in the park where the presence of the species was found. Every other day, between mid-July and mid-August, a researcher went to the study area to intercept the flight of the adults. They used the 'capture-marking-recapture' method to estimate the size of the population of these butterfly individuals, their density at the site, their survival and their possible tendency to dispersal.

The butterflies, captured with an entomological net, were marked with a non-toxic and indelible marker on the rear wing and finally released. Part of the study was also dedicated to the quality of the habitat and above all, to the state of conservation of the plant on which the butterfly depends. On a weekly basis, the researchers detected the various phases of the life cycle of *Gentiana pneumonanthe*, counting the number of buds present and the number of those suitable for laying, as well as the quantity of eggs laid, the number of holes present on the buds as proof of the entry of the larva inside them and then again the total number of ovaries that have reached maturity.

The movement capacity of this butterfly is rather limited and it is unlikely that it will be able to find other suitable places nearby should the conditions necessary for its survival fail. Climate change can pose a threat to *Phengaris alcon* and the particularly low spring and summer rainfall, like those just passed, seem to have had an influence, delaying the flowering of the gentians and creating a desynchronization with the butterfly. We can do little to remedy these aspects, while we can try to contribute to the conservation of this species by maintaining certain management practices that favor the meadows and avoid their abandonment.

THE CITIZEN SCIENCE EXPERIENCE

The educational mission of protected areas also pervades the world of research and management of environmental resources, including fauna. Thanks to 'Citizen Sciences', today thousands of people participate in the collection of data on the presence and distribution of birds, bees, butterflies, snakes, and dragonflies, contributing in a sensational way to knowledge as well as becoming more and more passionate about environmental issues. In some cases, this army of citizens collects and amplifies complaints from the scientific world, helping to change and create policies that are becoming increasingly urgent. One current example of this is the *European Butterfly Monitoring Scheme* in which *Butterfly Conservation Europe* and *Centre for Ecology & Hydrology* have collaborated to create an app and website where citizens can engage with the collection of scientific data, alongside learning about how to observe various species of butterfly (eBMS n.d.).

This civic action increases the possibility of collecting large quantities of data during visits to the park, transforming the experience of the park into one of an active visitor. The citizen not only gains in the soul refreshing vitality of nature but feeds back into the health and protection of that environment. When returning home, the individual

11 - Dragonfly. Stefano Torriani. *Libellula quadrimaculata*, 2015. (Illustration courtesy of artist)

may also continue exploring their surroundings, sharing observations, aided by a technology which allows multiple data and research centers to view it and use it. This amplification of data collecting is critical as many remote zones have otherwise limited capabilities of collection, yet we are all aware of the criticality of many species such as pollinators for the wellbeing of all species.

NO MORE LEAD!

Beyond simply educating, the dissemination of information, through group actions or activism, is important to help implement changes in legislation and the status quo. For example, the practice of hunting in the Alps has been under scrutiny from advocates for change, who are calling for change from both the hunters and the legislators. *No more lead for hunting. Stop intoxicating raptors in the Alps!* (2021) is a campaign which intends to promote the use of lead-free ammunition in hunting, the

12 - Bearded vulture, threatened by lead contained in ammunition used in hunting. Stefano Torriani. *Gypaetos barbatus*, 1993. (Illustration courtesy of artist)

presence of which is scientifically proven to be particularly harmful to some species of necrophagous birds. Vultures and eagles can feed on the remains of animals injured during hunting, or on innards abandoned in the wild, and fall ill with lead poisoning with lethal effects. Hunting could be practiced using bullets made of alternative non-toxic materials such as copper and steel which are now manufactured by all the major manufacturing companies worldwide. The solution is therefore to ban ammunition containing lead, but new laws and provisions are needed which are currently being examined by local, national, and international administrations. These need to be supported by everyone, including hunters. In turn, this campaign was launched in 2021, by local citizens, scientists, and activists, to collect signatures in support of these legislative changes (change.org 2021).

CONCLUSION

All parks' critical mission of maintaining or reintroducing a balance in biodiversity in these disturbed and still unrestored lands is carried out through several actions including research, education, and dissemination. The aim is that the comprehension of the 'others' as part of a system, rather than something to be controlled by humans, will evolve to allow all species to thrive, and for balance between species to be reestablished. This is a long process, difficult to achieve in protected areas and certainly rare in urban areas.

The wilderness of the Val Grande, as a chosen case study, offers opportunities to learn about a natural world that also invites human beings as visitors. Many modern humans have lost their natural dimension but still desire it, in order to free themselves from pollution, disease, and stress. The parks seek to satisfy the needs of nature whilst offering these positive experiences. There are no grand performances or survival tests, but opportunities to live authentically, embracing adventurous learning from the past to understand the reality of the present and reflect on the future.

The appreciation, caregiving, and rethinking of the relationship between humans and other-than-humans in nature can only originate from the sharing of experiences and from listening to the teachings of nature, which always tends dynamically towards new balances. Scientific research and projects (e.g. repopulating, tracking, monitoring) and the dissemination and engagement of local communities such as the 'Citizen Science' is an active part of the progress towards this new equilibrium. Through deep observation, reflection, discussion and positive action we are able to move closer to a system in which all living organisms thrive and are equal players.

Coordination with other governmental and policy making entities (such as parks, nations, neighbors, conventions, the EU) and also globally to mitigate the increase in human-made climate change, has been key in the creation and protection of these fragile spaces. We can see more clearly, from the vulnerability of habitats such as the Alps, how there is very little separation between the global and the local, when considering the cause and effect that leads to the destruction of our wilderness. It is now time to continue and increase the actions that restore and defend biodiversity.

Voices of the Fellows
Animals

Lucinda Coleman, Editor

Coexistence encourages an expanded consciousness of ecological reshaping through sensorial and reflective dialogues. Artistic, creative, and scientific explorations invite shifts from egocentric views towards a holistic approach where humankind lives in harmony with other species.

While such collaborative approaches are inevitably messy and unpredictable, there is an inherent acknowledgment of interdependence between all living and nonliving entities and the critical need for coexistence through site-making processes. The work of the fellows-in-residence is merely a beginning, as Patrick Jaimes has observed, "Artists can only hope to spark discussions, seed questions.

At the same time, some also strive to challenge systems and offer new possibilities to open doors. Artists—and authors in a broader sense—are by and of themselves limited in their reach but can and should be part of a broader movement".

The following voices resonate with the desire to share new ideas and ignite conversations about how animals are seen, understood, and represented.

1 - Horse at r3. (Image courtesy of I. Mazzoleni)

2 - Interactions between animals and artworks at NAH*R*, 2021. (Image courtesy of NAH*R*)

ON MORE-THAN-HUMAN ALLIANCES

Alliances of the past that could still exist today do not. Recognizing the harmful legacies of ownership and control helps me relearn embodied stories of our old, sacred dependence on the non-human.
Jonathan Pando Ocón, Geographer

I am interested in alliances of care, communication, and community in my work. How do humans care for animals, how do they communicate that care, and what kinds of communities are formed in the process?
Susan Haris, Multispecies Ethnographer

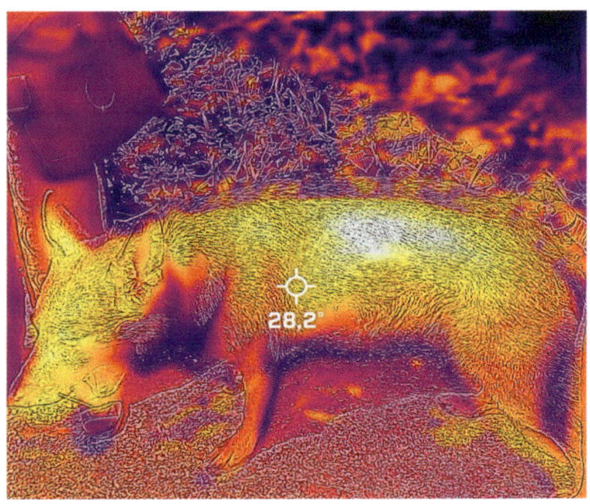

3 - Looking for other ways of seeing our kindred species. Pictured here is the same dog as in fig. 2, through thermal imaging. Petra Gruber. *Lilli*, NAHR EMPACT, 2023. (Image courtesy of architect)

While the search for horizontal alliances is what drives my practice, both this and the use and instrumentalization of sentient beings inform my work. This idea of alliance goes beyond the botched notion of 'humane' treatment. We don't really need to expect something in return from our relationship with animals, but even

when we don't seek to extract value -monetary or otherwise-, they are valuable in more ways than we may realize yet.
Patrick Jaimes, Artist, Photographer

Today we pursue the fetish of technology and forget the human and the need for harmony and coexistence among all species. The risks of extinction of certain species, and the loss of biodiversity are not unrelated to our own lives. In my artistic practice, especially since COVID-19, I have introduced this issue in my images.
Meri Gorni, Artist

I build my sculptures by collaborating closely with other species. From plants to insects, their construction methods, and their way of shaping the world are incorporated into my practice in a reciprocal exchange. I work mainly with salvaged materials and objects—'contemporary ruins'. Anthropic and non-anthropic elements, manipulated by a plurality of species, and together with all of them I build hybrid sculptural organisms where the shapes of wood are decided by woodworms and mosses, and crab claws are cleaned by wasps.
Camilla Alberti, Visual Artist

5 - Meri Gorni. *Nothing is as before*, 2021. (Image courtesy of artist)

4 - Wasps cleaning crab claws (L). Mosses & plants (R). Camilla Alberti. *Ruins*, 2019. (Images courtesy of artist)

6 - Karl Logge and Marta Romani. *A Weird Entanglement*, 2021. (Images courtesy of artists)

I am quite interested in the way human habitation and designed structures affect and interact with the native web of species and matter of a specified environment. Learning from and aligning design decisions with the members of an ecological community will strengthen our relationships with more-than-human beings.
Emma O'Mara, Architectural Designer

Ancient weaving is a lot about more than human alliances. In many different cultures, since time immemorial and surely in the tradition we are tied to, designs and patterns carry animal and plant knowledge or territory-specific understandings, mythical and symbolic meanings, and attunement to natural phenomena and cyclical codes. But they also present other more-than-human figures or 'abstract' shapes that belong to different zones of the memory, and they witness important alliances between land, sea, sky and all the threads of life that inhabit these spaces. In this way, our practice explores the sustainment of endangered multi-species modes of being that are fundamental and strong.
Marta Romani, Artist

The intimate and introspective practice of weaving that we practice includes moving throughout the island [Sardinia]. This embodied mode of practice informs an ongoing process of working with various non-human subjects and collaborators including different plants and seaweeds, Australian native bees, eels, various breeds of sheep, and migratory birds. Whilst more specifically looking to bring forward issues related to post-urban and rural futures, artistic and scientific collaborations, and the relation between territory, meaning, and memory through thinking about things like nets, traps, waterways, looming, and the ontological qualities of this dance that shapes the human as this weird animal that is able to see itself as other-than-the-animal.
Karl Logge, Artist

I am especially interested in the half-human, half-animal gods in Hinduism, for example, Kamadhenu. In Hinduism, cows symbolize mothers and feeding human beings. Animals in myths show the close connection and observation between primitive humans and animals. Humans depend on animals, such as for consuming milk, so they learn to respect and coexist with animals.
Xinning Liu, Animator

7 - Interactions between animals and artworks at NAHR, 2021. (Image courtesy of NAHR)

ON RESTORING BALANCE BETWEEN SPECIES

I believe that inquiry-based endeavors may inspire a movement towards restoration of balance due to the highly personal and creative nature of such exploration, and the sense of community to be gained through sharing these inquiries and reflections.
Emma O'Mara, Architectural Designer

8 - Sculptures of two calves made with locally found materials. Yulia Shtern. *Carmen and Gloria*, 2019. (Image courtesy of artist)

I never want my work to be didactic. However, I do want to have some sort of an influence where the work creates a space for deeper contemplation on our relationship with the nonhumans. For me, a place to start is to ask and raise questions and create work that challenges these ideas and relationships. Human and non-human female bodies are associated with giving life and nurturing. But non-human female bodies are also 'objects' of exploitation in the realm of providing food and driving the economy. My performance work at NAH*R* with dairy cows focused on the idea of exploiting female bodies. I placed my body in the environment of the dairy cows and grieved and mourned alongside them.
Laleh Motlagh, Performance Artist

9 - Laleh Motlagh. *Love in Cheese*, 2021. (Image courtesy of artist)

A balance insinuates a removal of challenges and conflict, while restoration insinuates an ideal that previously existed. Neither are realistic, nor should they be the goal. Coexistence is messy, but to move from an 'ego' to an 'ecos' ethos, we must humbly recognize we are an exception. No other species has had such a prolonged and continuous impact on the world, but in collaboration with the (non)human, we can relearn our connection to all species.
Jonathan Pando Ocón, Geographer

My work blurs the line between man and cow. Humans living in cities are almost separated from the animal world, treating animals as second-class creatures. I hope to rethink the fundamental difference between humans and animals by merging humans and cattle. We all have blood, flesh, bone, and skin. Humans, as social animals, retain a lot of animal nature. This theme is evident in animations, especially those in which anthropomorphic animals are the main characters. Those hybrid characters show something that cannot be done in human society, but something that humans can dream of doing. In my animation, I had complete freedom to express myself in the form of the species, and feel I created something in the space beyond the animal and human divide.
Xinning Liu, Animator

10 - Xinning Liu. *A Theatre in Taleggio*, 2021. (Image stills courtesy of artist)

11 - Meri Gorni. *Nothing is as before (Niente è come prima)*, 2021. (Image courtesy of artist)

12 - Francesco Maluta. *Woodpecker*, 2021. (Image courtesy of G. Gritti)

I try, in my images, to ask questions. Our resources are finite. And the way we live is not a one-and-done world, so I think we need to redesign the relationships between species even if I don't have much faith in a society that sees us only as buyers and consumers. I understand, however, that we need to heal the wound between Humans and Nature as soon as possible, but I don't know how much Art is able to move the needle.
Meri Gorni, Artist

ON OUR ENTANGLEMENTS

I am suspicious of hopes for such a balance because I don't think all humans could ever be part of such a democracy when they are struggling to survive. My endeavor is to try to get at those people who nevertheless dream of such a zoopolis: animal rights activists who relentlessly tilt the balance in favor of animals.
Susan Haris, Multispecies Ethnographer

...in my practice of temporality, expectations are punctuated by the rhythms of the *companion species*, as defined by Donna J. Haraway, I constantly collaborate with plant organisms, fungi, and insects that inhabit my sculptures or help me in their creation. It is a constant exercise that helps shape a certain attention towards who and what surrounds us, recognizing their style and times of life. An almost votive dynamic that ignites my attention and forces me to be in the process, to inhabit that relational moment. A rhythm that I diffuse and narrate in my work.
Camilla Alberti, Visual Artist

14 - Camilla Alberti. *Plants Architecture*, 2021. (Image courtesy of artist)

13 - Portrait of an unrepresented working animal, the donkey. Nora Sweeney, 2021. (Image courtesy of artist)

15 - Camilla Alberti. *Plants Architecture*, 2021. (Image courtesy of artist)

16 - Karl Logge and Marta Romani. *A Weird Entanglement*, 2021. (Image courtesy of artists)

In Val Taleggio we used this embodied and performative mode of investigation to rethink and reconsider remnant buildings and towers known as the Roccoli. These once sat within an elaborate system of trees, traps, nets, and birds, and through the art of weaving, we interrogate these as potent sites that trace different threads of life, memory, and knowledge, revealing a weird tapestry and hybrid ecology that might offer other ways of looking at inter-species alliance.
Karl Logge, Artist

When we visit a new territory, we are really alert and sensitive to certain landmarks, sounds, events, and encounters. They emit special signals that cannot be ignored but at the same time are difficult to pin down. We found the zone around Sottochiesa rich with these clues and cues, possibly because of the intense presence of water. And here comes the responsibility: when the essence of these places emerges, it is usually because the moment is critical. This means that a sense of imbalance is there to alert us about something important. It says wake up, make haste, and be quick to love each other! So, we try to pick up the signals, gather, connect, transform, rework, and embed them in our artwork to tell a new story. We hope our art inspires curiosity. That's the first step to any love-based relationship.
Marta Romani, Artist

17 - Creating encounters through becoming signals. *Coexistence* exhibition in Bergamo City, Italy, curated by NAHR. Sophia Aitken. *Pollen*, 2022. (Image courtesy of artist)

To entangle is to wear the work *Headnet* whilst embedded within the chorus of birdsong, this is to remove yourself from this world when standing within it, to become a thing amongst a network of many things—to become conscious yet also let go of thinking too much, and therefore move beyond the idea that the human stands as the central issue, problem or resolver of the times we are all living amidst and that there are perhaps much bigger forces at play, and the future could be weirder than we can possibly imagine.
Marta Romani & Karl Logge, Artists

19 - Nature and culture in coexistence. Fiorella Pesenti Campagnoni, 2023. (Images courtesy of F. Pesenti Campagnoni)

18 - Karl Logge and Marta Romani. *A Weird Entanglement*, 2021. (Images courtesy of artists)

ON INTERCONNECTEDNESS

First to specify my sense of what ecological architecture entails, I am imagining a design process that prioritizes creative and contextual goals over aesthetic production. A process that takes extended time, research, and care towards existing ecological networks before altering the physical and invisible space. Ecological architecture entails localized materials, whose tactility and history will offer a sense of place to inhabitants. To me it seems that these endeavors toward ecological community building will result in spaces more entangled to their environments, providing shelter, not separation. These invigorating and unique ecological atmospheres invite our senses to participate, rather than to be cut off within insular walls.
Emma O'Mara, Architectural Designer

We have always moved as if we were not part of nature! The experience of the pandemic [COVID-19] has taught us our fragility and vulnerability. Have we realized the conceit with which we felt and believed ourselves to be the center of the universe? Have we understood the relationship that binds us to one another? Have we understood how little it takes to disrupt the balance of coexistence and co-existence between species? What if all this carelessness leads to the disappearance of the human? Our presence is not so indispensable for the birds to sing and the grass to grow!
Meri Gorni, Artist

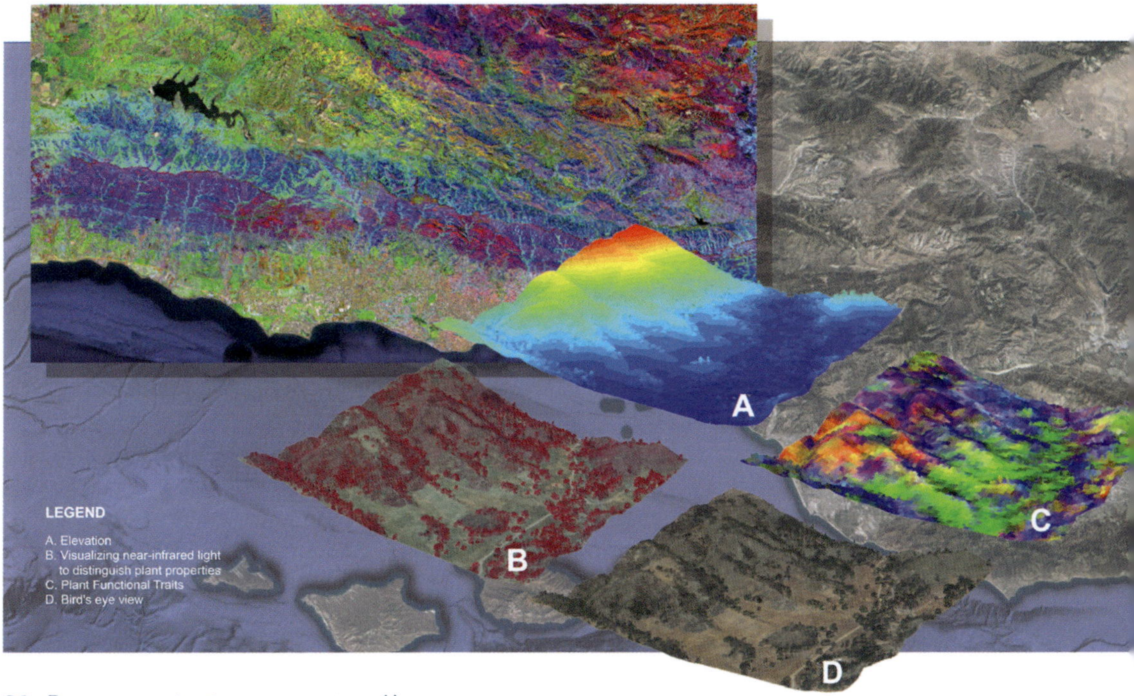

20 - Remote sensing imagery captured by satellites, planes, or drones help us observe the (non)human world from a perspective that borders science and art (SHIFT 2021; National Agriculture Imagery Programme 2020)[1]. Our interpretations of this data trace landscape changes across the (un)natural world and discover connections between species. Jonathan Pando Ocón. *Eye in the Sky*, 2023. (Image courtesy of geographer)

Depending on the method of image capture, remote sensing can track landscape changes, current conditions, and at very high spatial resolutions, can even detect presence such as a bird nest in a tree. This leads humans to questions of causality and impact on a multi-species level

Jonathan Pando Ocón, Geographer

21 - Meri Gorni. *Altre Voci del Verbo Parlare*, 2020. (Image courtesy of artist)

[1] The Surface Biology and Geology High-Frequency Time Series campaign (SHIFT) was an airborne and field research effort jointly led by The Nature Conservancy, UCSB, and NASA Jet Propulsion Laboratory (JPL) in 2021. Airborne imagery for SHIFT was captured using the Next Generation Airborne Visible/Infrared Imaging Spectrometer operated by NASA JPL.

ON ETHICS FOR A MORE-THAN-HUMAN WORLD

Historically, landscape, a term that emerged in the 15th and early 16th century is a deeply humanist idea and profession. The animals' viewpoints reveal an alternative way of looking and working with landscapes- shifting from the traditional landscape way of seeing the world, one that organizes around the composition of space and the perspective view of a detached individual outside of the landscape, to a practice that consists of active engagement with a living world of which all things (plants, animals, humans) are a part. My project offers a novel vision of Val Taleggio. It reveals connections and processes that lie latent in the landscape—invisible to us but intimately palpable to frogs and their amphibian relatives.
Tami Banh, Landscape Architect

23 - Patrick Jaimes. *Mose*, 2021. (Image courtesy of artist)

I find it challenging to let the *animal* 'speak', not only because of the obvious reasons and theories but also because the animal does not care to speak to us. This indifference displaces the paternalistic welfarism of the animal lover's *LOVE* and *CARE*, and leaves a strange void in its place. Multispecies ethnographic methods try to be radical enough to respond to this question and not just let the human go on articulating.
Susan Haris, Multispecies Ethnographer

Animal ethics has yet to offer incalculable value. As has been documented, many forms of supremacism start with the marginalization -and sometimes animalization- of the 'lesser' other. As this thinking expands, and breeds power and excess, ethics may prove critical going forward in a world in crisis.
Patrick Jaimes, Artist, Photographer

22 - Tami Banh. *Seeing Through the Frog Eyes*, 2021. (Image courtesy of artist)

25 - Patrick Jaimes. *Cow M*, 2021. (Image courtesy of artist)

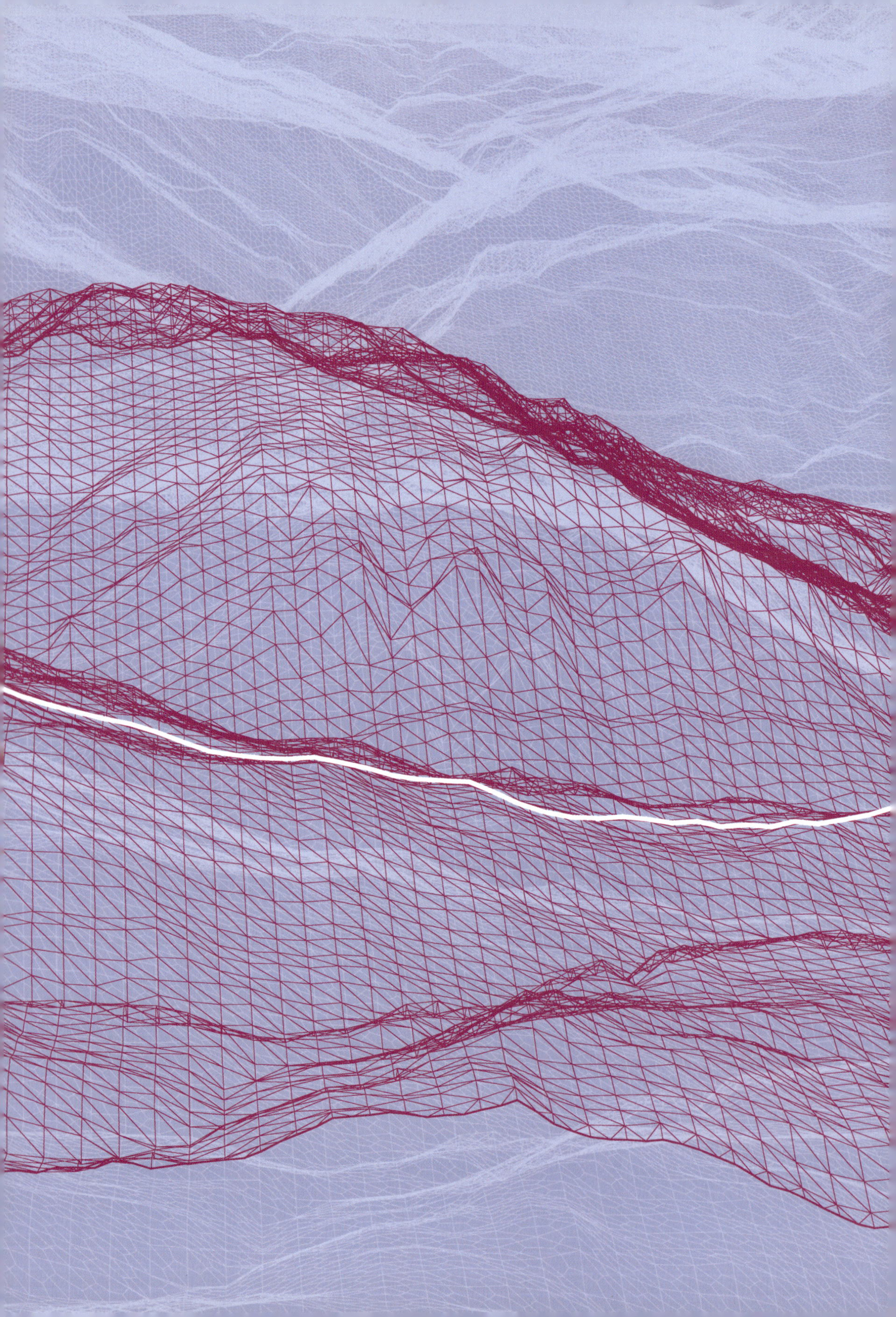

CHAPTER 6

Soil

The Critical Zone

Introduction

Richard Molina

Although it may seem basic, there is nothing simple about soil!
It is a complex symbiotic matrix of life that nourishes countless inhabitants.

When we think about soil, our first thoughts can often be that it is just dirt. We might think it's dirty, infested, in the way, or that we own it. So, we remove it, displace it, grade it, cover it, wash it off, mix it, and divide it; all this without considering the inhabitants we inattentively irritate or displace. Insomuch that more than 75% of the Earth's soil is gravely depleted, while the remaining 25% is usable for cultivation (Montanarella, Scholes & Brainich, 2018). We routinely overlook the vast array of soils used in our everyday lives from the backyard we sit in, the trails we hike with our dogs, to the food we eat. Soil is everywhere, and the worlds of soil are tremendous! And we cannot undermine our role as inhabitants, moreover designers, in establishing a balance within such a delicate ecosystem.

We are part of it, as much as it is a part of us.

It feels invigorating when we uncover positive revelations from the deep and densely populated network of life found between the roots of trees, plants, and grasses where organisms, microorganisms, natural filters, and water sources all thrive. Soil science analyzes and provides collections of data that can help us understand optimal soil conditions for our needs in everyday life: from developing infrastructure, assembling non-traditional biomaterials, and cultivating vast farmlands, to developing climate-proof crops. The data also informs us that we are the prominent factor that gravely influences the internal transformations occurring in the labyrinth of soil - impacted by direct contact of foreign elements at the surface level.

Interconnected and influenced by all those elements that surround it, the tree is a single byproduct of the soil, but the soil relies on the tree for the necessity of the shade produced by the canopy through photosynthesis and the carbon-rich sugary root exudates passed to the microbes in the soil that feed the cycle of life. By observing this process, we can trace every molecule, sediment, and foliage that is relinquished by each of these independent systems working to sustain one another to create just the right conditions to continue its evolution and protect itself from the harmful effects of climate change. But understand that this is only a minute speck of Mother Earth's 13.00 million hectares of land (Ritchie & Roser 2019). It only takes subtle changes to transform the landscape beneath our toes, and the ecosystem of which we are a part, to potentially erase species and introduce new invasive species into territories where they had not been present before. It is telling that we have enough imprinted knowledge and technological expertise to predict the effects of such repercussions.

In the unfolding chapter, both Marilyn Poitras and Alessandro Longhi describe various instances in the history of soil through the lenses of natural order and science, respectively. Poitras delves into several creation stories and Indigenous Cree depictions of Mother Earth to awaken us, reveal our position within the circle of regeneration, uncover the complex understandings and network relations of soil, and renegotiate our feudalist histories to reestablish our relationship with Her, as Mother. Through the lens of science and data collection, Longhi exposes the interconnected network of biotic and abiotic inhabitants, within the atmosphere, hydrosphere, biosphere, and lithosphere. Both authors distinctly trace the origins of soil and define its role as an independent and interdependent matrix of life, which has been disrupted, disturbed, pulled apart, subdivided, and owned as though it is a commodity that exists only to serve us.

In weighing the scales of imbalance, humans are the counterbalance. As such we are necessary to complete the cycle and reconsider our interconnected relationship with our environment to achieve homeostasis. This is exactly the relationship that the fellows-in-residence, in both the Taleggio Valley and Santa Ynez, have explored under the prompt, *Soil: The Critical Zone*. In the following chapter, the Fellows reveal distinct relations with soil which are accentuated in their research, art, and discoveries.

As both Longhi and Poitras point out, soil is a memory-keeper that stores information and stories in a multitude of layers. As future generations continue to dig into and dissect these layers, what rich information will they unravel from the soil and its critical zone? Will it be composed of artificial matter suffocating microorganisms or chemical byproducts leaked into lands and habitats by agricultural and industrial zones? How do you want to be imprinted into the memory of soil?

As inhabitants, and as part of the soil, we must consider our direct impact on the soil and modify our general approach. As designers, we need to encourage and reestablish environments where landscapes are not bypassed as a secondary or residual element, but rather a crucially interdependent aspect of the design ideology that integrates more conscious parameters to counterbalance positive and negative impacts on our artificial and natural environments. Every line we draw, every material we select, and every surface we impose on the *terra* impacts a delicate ecological infrastructure, which further reforms the surrounding lands to support the city through agriculture, mineral extraction, infrastructure, etc.

Not only one solution exists to regenerate soil and bring it to the forefront of our thoughts. The myriad of solutions require a counterbalance, and we are vital in the collaborative interplay with non-living elements and the multitude of living organisms in the soils beneath our feet. Again, there is nothing simple about soil.

Mother Earth: Coming Home

Marilyn Poitras

What are the stories of your people or peoples? Stories enlighten our lives and promote reconciliation between peoples, and all the elements of our shared earth. As Elder Gladys Wapass-Greyeyes has noted, "we are the smallest part of creation, dependent on all else for our very existence" (2012, pers. comm.). During the *Truth and Reconciliation Commission* in Canada, Chief Commissioner Murray Sinclair identified an important set of questions: Where do we come from? Who are we? Why are we here? Where are we going? (Anderson 2016). These four questions frame the reflections of this essay for they apply to us as individuals and as a species on the planet and connect us to one another. As Sinclair continues to remind us, we are all on this quest together for an understanding of our origins and interconnectedness.

The earth beneath our feet holds us here and is our home, our refuge, our collective ecosystem. Traditional knowledges and the ancient teachings of life have been dependent on an understanding of the interconnectedness of humans and lands. Our stewardship of Her, the earth, our Mother, held those teachings. Folded into ancient creation stories, the stories taught not only what to take of Her and how to live within Her for balance and sustainability but also, what sustains Her, and how each and every component of the life on top and within and above Her plays a role in that relationship.

Yet, I believe we have forgotten our relationship with the earth, our Mother.

1 - Mother Earth/Earthrise. Lunar Reconnaissance Orbiter, 2015. (Image courtesy of Visible Earth, NASA)

CREATION STORIES: *WHERE DO WE COME FROM?*

Creation stories are the means by which many communities have come to codify, or organize, our interrelatedness, our interdependency with our Mother, Earth. They are alive and accessible in Indigenous cultures and/or amongst original peoples. These stories matter as much as the particles of organic mass, gas and liquid that makes up our soil.

The stories on how we 'come to be' are as fantastical as a god creating the earth in seven days, of fallen powerful women and condemning powerful men, of angry sons, and enticing daughters, and of demigods, and goddesses ruling the heavens and the netherworld. Still others tell of humans pulled from clay, from the soil itself, and some that teach that we were brought from the stars. We are the stories we tell ourselves:

> The stories about the time following creation are filled with metaphorical tales about transgressions in the so-called ecological compact. These myths, stories of creation, stories of origin and of emergence are similar in the sense that they are told by Indigenous people all over the world and all describe compacts built on relationships established between humans and other living things. These stories embody the understanding that humans, along with all other entities, continually create the world (Cajete 2000, p. 78).

Ancestral connection to land, in the sense of the actual soil upon which we rely for our very lives, is very nearly a distant memory as the earth's life force and all Her riches are extracted. Arguably, little is known of what that extraction does to the

2 - Forest soil samples, felting, and embroidery. Ñemikafe. *Suolo della foresta*, 2022. (Image courtesy of artist)

soil and its rich and varied internal ecosystems, or to the plant and animal life on the top. Little is shared about the practice of subcutaneous injections for the purpose of extractions of every kind. Instead, we parse her out. She is partitioned, dissected, divided and disemboweled: an act resulting in a constant cutting of ourselves off from Her, and She from us.

The typical way we discuss the dirt beneath our feet is by ownership. Ownership through the continent, the country, the state, province, district, city, village, town and then often by the person whose name is attached to the land. Laws are then created to limit access to the land by other humans and in the case of animal husbandry, limits to the animals themselves. This view of land through ownership is common, and the act of mapping alone could absorb libraries of information. This type of labeling, of categorizing, is the way we measure, attribute, even expand power over the domain of soil. Occupation through flag and blood is the action within the pages of recorded human history used to measure the strength of a nation. Conquer the people and take the land; the more the better. Boundaries are the consequence of this act of ownership, control, and domination. Extraction of the riches of the day are the aim; this is the act of colonization and commodification of the earth Herself. This has become our relationship with soil.

The categorization of land 'occupied' is thus colonized. Imperialism stretches through the hierarchy of nationhood, and is then further subdivided into private and public spheres of control. The value attached to each inch, mile, or hectare is dependent on who is doing the calculations of the profitability of her component parts. A farmer may measure the richness of the dirt and its ability to produce the crop or animal to be husbanded. A military general will look at this from a strategic operational lens. A developer designing a multi-home community might look from another lens altogether. We can then look at the urban and rural use and commodification of land, and/or whether citizens can grow gardens on the walkways or raise chickens in their backyards, all through the lens of ownership, title, and control. We do not look through the lens of Her as a life-giver.

The colonial tactic has been to take land based on an imagined notion of the land being unoccupied. The doctrine of *terra nullius* itself is worth some study for our relationship with the earth. The Latin term, *terra nullius*, refers to land legally declared unoccupied by peoples and has been used to justify invasion and colonization. The British government used the term to justify the invasion of the continent now called Australia, despite the territory being cared for by over 500 First Nations people groups at the time of colonization. While the *Mabo* case successfully overturned the myth of *terra nullius* (AIATSIS n.d.), the devastating consequences of colonization are still evidenced in the lives of Aboriginal and Torres Strait Islander peoples, and within the landscapes of Country today. Our *terra nullius* debate does not even brush up against any notion of the soil itself having independence, or that we are interdependent and in relationship with Her. Shall we also disrespect the laws of Nature? Or shall we listen? Our creation stories will hold those teachings.

There are parallels to be seen in an ongoing relationship with soil as an unoccupied space to be taken up for wealth gain, resource extraction, and development. The violations of the soil itself are less likely among things measured, recorded, or noted. Although the ill health of many species living on Her provides that record in other ways.

Close your eyes and see the landscapes of your homelands: highways, fences, roads, lines of latitude and longitude, invisible borders, flags, powerlines. There are so many ways to show that we own, control, possess the land. In Saskatchewan, Canada, the entire geography is divided into square mile grids. Because the province is not perfectly square, this is not a fluid system and to compensate for that there are 'correction' lines. This way of dividing land was the practice of the dominion lands survey in the colonization of the lands in the western portion of what is now Canada: parceling the land into contiguous patches for distribution is the act of colonizing, owning, controlling land. The new settlers arrived in swarms then overtook the lands in one-mile square allocations. There were imagined boundaries between people, crops, animals, and soils, and imagined control as well.

3 - Subdivided land - agricultural land use in the UK, 2023. (Image courtesy of I. Mazzoleni)

In English law, the perpetuation of the idea of control is based on the Crown holding underlying title and thus estates can be further subdivided for the use and occupancy of the land, at the discretion of the Crown. The ability to retain land control within a family who captured an estate was found through an interest of fee tail in land. The original practice of only male issue, wherein title could pass to heirs and issue boys, changed in 1922 with the *Law of Property Bill*. Multiple misogynistic metaphors are hard to overlook in this history through this lens of our relationship with the earth. It plays out on multiple fronts. Feudalism and the feudal pyramid arose out of chaos in Europe post-Roman Empire to create a system of a master-owner and subinfeudation, or smaller parcel occupiers. There may also have been other people who were on the land and connected to it, sowing in it, eating from and working upon the land. There were most certainly other species doing so.

Breaking down the demand for profitability further we can examine the use of chemicals in the food industry, harvesting practices, our waste practices, and our extraction practices. Chemicals are used to force further production, to kill dependent plants and insects; even the seeds themselves. We can also examine how our own human medical industrial waste is wreaking havoc (Buhner 2002). We can see our expansive human footprint on soil when we look more deeply at the microorganisms and how they are impacted or eliminated. We can see our past through the human lived story interwoven into the soil and we can predict our future from it. Mother Earth is a historian also and can tell the tales of the lives She supports us at the cost of loss of connection to Her needs. How might we pick up that thread of connection to re-establish our interdependency? How can we rewrite our story and read it in the soil in the future?

Myth or mystery, fable or legend, creation stories are essential to our relationship with soil because they remind us of our connection to the earth and all that exists within, upon, and outside of Her. There are many ways to know Her, to know ourselves and find our own understanding about how we come to live on Her, to see how we

4 - Earth as mother. Anju-Marie Kasturiraj. *Le Madri*, 2022. (Image still courtesy of filmmaker)

live through Her and, ultimately, to know how we die with Her. That is the cycle; it is the reality of all that exists here and we can see ourselves within that if we can understand our relationship to earth, the planet and within this discussion, the ground itself. We can remember, reconnect, and repair with her.

5 - Web of Connectedness, 2023. (Image courtesy of I. Mazzoleni)

Roles and responsibilities are often found within creation stories, and within them we will find ourselves as villains, heroes, friends, and foe. For example, the North American Indigenous people, the *nêhiyaw*, also known as Cree, tell the creation story of the Star Woman, *achakosak iskwew* and her travels through the sky from the stars to land on the back of a turtle, leaving marks related to the moon cycles.[1] The story gives us details about the birds who caught her fall and the struggles of the animals near her who aided her by going to the depths of the water, under the turtle, to locate some earth to use, to create a place for her to live. It is a love story of how so many animals perished to find that earth for her and ultimately how the tiny able muskrat, near death, is the hero who returns with earth in his claws to give to *achakosak iskwew*. The Star Woman is then able to expand that earth to create a larger island for all beings to live upon.

Everyone has a role within this *nêhiyaw* story: each bird and animal described in the full version are working to bring her soil so that *achakosak iskwew* can live, eat, replenish, and contribute to their lives too. Each creature has certain characteristics that reflect their gifts and therefore their contribution to this place. The woman is the life giver and expands the earth for all. There is even a component for the spider, *apihkêsîs*, who ultimately symbolizes the web of our connectedness. Everything is connected: the sun to light, the moon, the stars and the flyers, the earth and the animal kingdom, the plant kingdom, and the crawlers, and the water spirit and swimmers.

Whether it is through Romulus and Remus, Zeus, Hwangun, Rod, Ometecuhtli, Pan Gu, or thousands of others, there are a multitude of creation stories to find within every culture. Some tell of the birth through creation of children from one being, or of the egg that produced the hero, or of humans created through pure will. Others crossbred animals and gods and humans. All have a connection to land, plants, and animals. These connections are the key and they vary from life-producing to life-taking. The stories tell of power and the dynamic relatedness of all beings within the story, upon the earth, or from the heavens, or the depths of the universe. Creation

[1] This is an oral story passed down through families and within communities but there are written versions of it as well. https://www.historymuseum.ca/cmc/exhibitions/aborig/fp/fpz2f22e.html

stories nourish your soul, connect you to your ancestral lands and awaken your blood memory of who you are and why you are here. As Leroy Little Bear has observed: If you look at things from a flux point of view, a geodesic spider web in motion, you begin to see that everything is related. Whereas in Western thought we try to isolate, we try to get down small and smaller and smaller and isolate. Whereas in Native thought it's always holistic thinking, it's always the relational networks (BanffEvents 2016).

6 - Angus Rushin and Lily Garget. *Coexisting with Humans and Trees*, 2022. (Images courtesy of artists)

Leroy Little Bear is Blackfoot but his relatives, the *nehiyaw*, see things in a similar light. Where does life come from and how is it maintained? It is a story that cannot ignore birth and seed, sacred and sacrificed, and soil. For most Indigenous peoples in North America, the earth is our Mother: the soil is Her form, the waters Her milk, the grass Her hair. She works with the metaphysical to keep this planet spinning, growing, and shifting. Everything upon Her is animate and every being plays a role: wind, thunder, lightning, cold, warmth, rain, frost, heat and fire are all part of this story. Decomposition, worms, mitochondria have main character roles. It is an interdependence dance we can only imagine. We are reliant on them coalescing, each relating and interacting. It is a symphony: our symphony. The natural elements help the Mother as She cannot do it alone. This all works together. There is always sacrifice, celebration, abundance, and loss. There is never a disconnection of spirit. There is language and dance in this relationship of all the elements filtered through the soil. As the *nêhiyaw* Elders have reflected:

In each Cree territory, covenants were established with land, water, plant-life, animals, etc., which would act as spiritual agreements of reciprocity dictating how all life would co-exist in mutual protection, benefit, and continuity. This reciprocity is maintained through the various ceremonies, songs, narratives, and personal sacrifices of Cree spirituality (Ermine).

CENTERING: *WHO ARE WE?*

The *nêhiyaw* lens or story shows us we are all related. Through another creation story these components or characters of rock, soil, water, animals and so forth, may be foes, competitors, or they may be lovers. Your creation story gives you an imagined frame for which to see yourself in your own history and therefore, in the present. Here is an invitation for you to re-embody your creation story and find your character within the plot of your life, your work, and your relationship with the Earth. What you have been taught about where you come from informs your worldview and may even show you a path to rekindle a relationship with the soil that provides the food for all creatures, to empower and inform your life choices. Specifically, how might we build a relationship with displaced soils? Are we willing to take responsibility for Her condition? What is our response to Her when our life force is in jeopardy?

It is said that the Earth we live upon is, Herself, a history keeper. She is also a mirror of our existence for She reflects us back to ourselves. How we look after Her, our Mother, is how we look after each other. Our study of Her can only ever be a study of ourselves and how long or how short our recent revolving turn on this planet has taken us. She is understudied, underappreciated, taken for granted, colonized, commodified and She is violated. Parts of Her are now barren, yet parts of Her are as rich to us as the gold deposits we mine. Soil is our black gold, our riches beneath our feet. How will we fight to protect the soil? How might we connect to the ruins we have made, and heal together?

As Mother, She represents every single relationship that we know to exist and those teachings must be so because She is our Mother. We know and measure the heave of Her tectonic bosom and She knows and informs our most shallow breath. How we live upon and within Her, it is said by our Elders, is how we will become the fertilizer for what will grow next. Reaping what you sow takes on another layer of earth in this context. Are you sowing for the future, for wealth, for revenge, or for yourself?

What can be reaped from this intercourse? Reaped for Her and for us? What are we birthing and what is dying in our lifetimes?

Understanding our relationship with Her is to know ourselves intimately. It is then we can create with Her, live with Her, grow with Her. Without an understanding of our responsibilities we only colonize and disembody. We objectify and become the villains of the story. The *nêhiyaw* creation stories teach us that what we do to the earth we do to ourselves, so She must be us and we must be Her. Seen through this mirrored lens, through this story context, we must interrogate ourselves:

› Where are our oil deposits, our underground springs, our microbes, and our root systems?
› Where do we adapt in life and in our bodies and where are we rigid?
› Where do we decompose (disease) and where do we evolve (immunity)?
› How are all the systems of our bodies interrelated and perhaps even intrinsically dependent, interdependent, or even codependent?

7 - Clay found in the Val Taleggio forests, 2023. (Image courtesy I. Mazzoleni)

8 - Anju-Marie Kasturiraj. *Border of Soil and Skin*, 2022. (Image courtesy of artist)

› What are our bodily habitats and patterns and boundaries; where are we rich and where are we contaminated?

We take from the earth all the time: food, nutrients, teachings, shelter. Taking requires giving; healthy relations must be reciprocal. Without reciprocity there is depletion, a certain ending. Wells go dry, nutrients are lost, crops perish, sinkholes develop, and creatures become endangered and then extinct. Life is unsustainable in one direction. We flatline: the death symbol for time itself. Where are the circles? The circles that allow for seasons and cycles and growth and relearning and re-teaching?

CIRCLES: *WHY ARE WE HERE?*

Circles are a very important symbol. As a pattern, circles spiral around and around to expose and to understand, constructing new shapes for use to interpret and to understand. Moving out of flat lines and triangular hierarchies to an equal playing field with no corners to hide in, circles are an inclusive shape. Circles are the shape of the sun, moon and planets; the teaching of a tree trunk, a bubble, mushroom fairy rings, eyes, or irises, flower centers, insect markings, and our own heads. Elders teach about circles and cycles always using visual properties which teaches to invite all the senses to participate in the understanding of what is being offered to us through our connection to nature. The offerings are of course the lessons on how to live here with Her, upon Her.

> As one Navajo elder states "you must ask permission of the plant or the medicine will not work. Plants are alive: you must give them a good talk" (Huges 1983). Such offerings are made to ensure that the pulling of the "Earth Mother's hairs" does not hurt her too much and so that she understands that you comprehend your relationship to her and what she is giving you through a part of her body. In honoring and understanding this relationship, people also understand their reciprocal relationship to all of life and nature (Cajete 2000).

When you are taught how to pick plants with *nêhiyaw* traditionalists, our herbalists, our Elders, you are sat in front of the plant and told to introduce yourself to it. If you have been given a traditional name, you use it in this introduction. You are taking. To take you give; first an offering of another plant, the tobacco plant, who has an agreement to be the conduit for taking messages to the spirit world for humans. Once

9 - Circles composed of soil. Sophia Aitken. *Soil Mandala*, 2022. (Image courtesy of artist)

you pick what you need, ensuring there is ample resource left for others reliant on it—no matter their size or shape, bugs, four legged beings, the crawlers below—you use it for yourself and gift it to others. Taking what you need means you are aware of what you do need, for what and why, and perhaps more importantly, that if *you* need, then others must need too and this teaches you to then find your place within that circle. Again, a creation story will give you the context and the imaginings for the foundations of this relationship and the remedies for its repair.

Through the complexities of the earth, the soil is again reflecting back teachings to us in what She offers or acquiesces for us to take. What did it take for Her to grow it? What are the perfect conditions for Her birthing; what strains it? Who are Her cosmological, metaphysical helpers? Those helpers must ensure seeds are not only grown through the plants but exposed and dropped back to earth through wind, water, birds, bugs, animals and fire. She does not birth Her children in rows or alone. She births in unison with the other beings around the seedling, the egg, the animal, that they can in turn feed on, or be fed to. She has Her own science and intelligence. She can also be merciless in the execution of what She knows are Her cycles.

Each component of soil has function and need. Disturbing those systems results in change and reorganizing. As the Mother does her dance of life, She is clear about her laws and what will and will not work. She offers those laws to us in Her every production. All Her brood offer or take life. She favors none of us and loves all of us. Humans are the smallest part of creation, dependent upon all else to survive. We are here to survive, to reproduce the species, just like all the other beings here at home.

CONCLUSION: *WHERE ARE WE GOING?*

If you ask yourself where you come from, and who you are, and why you are here and where you are going, you must come to an understanding of you, your species, and in the context of a greater system. You in fact bring yourself home. In this way of self-reflection through the traditionalist's teachings or the lessons from a multitude of creation stories planted and spurting all over the globe, the modern philosophers and ecologists arrive at the same page, or at least chapter. *Where do I go from here* is the birthplace for the eco-philosophic idea of ecological humanism:

> Ecological Humanism...calls for the appropriation of humans to Nature. We have to see the human as a part of a larger scheme of things: of Nature and Cosmos. We have to transcend and abolish the idea of the Protean (and Faustian) human. The consequences of this reversal are quite far-reaching, and I will just touch on some of them. On a more practical level, Ecological Humanism signifies, among other things, frugality, recycling, the reverence for Nature, which are really three different aspects of the same thing (Skolimowski 1990).

Mother Earth is patient. She will always be here. She is our home. We are the smallest part of creation. Yet in contrast, there are responsibilities for even the most dependent on Her to enter into this relationship with soil. There are laws: to take only what we need, to understand the needs of others, to see the symbiosis of the interrelationship of all of us. Reciprocity is the main tenant of this relationship, the judge if you will. What do you offer in exchange for *how* you take from Her?

How are our human or country laws set up in accordance with our relationship with Her natural laws? If they are not, we can examine how and where those disruptions are. If they are, we can see if there is a symbiosis. We must ask, what are our cycles? What are Hers?

What is the story we have created and what story might we tell or remember? Our myths and legends are much more than fables for entertainment; they are the umbilical chords to our connection to our Mother, the soil upon which we rely, and for which we have assigned responsibilities. The characters within the stories we create, we tell, we live, come in every size and with every imaginable need and offering.

Imprinted Memory - Sustaining Life

Alessandro Longhi

Fundamentally, soil exists as an independent body, with its own specific origin, properties, and morphology. However, throughout history, many human societies have understood soil through its utility in agriculture and been driven by the fear of its degradation. Soils are also memory-keepers whose constantly shifting components beg further investigation and understanding.

1 - A natural soil profile opened in Val Cavargna, Italy. 2016. (Image courtesy of author)

In the quest to understand and study soils, it is important to consider the effects of our actions on the various biotic and abiotic factors of soil. In order to reflect upon the state of our soils today we should begin by reflecting on the historical development of pedology: the discipline of soil. By considering the notion of imprinted memory of soil, we can learn about the development and state of the soil sciences. Then, having understood the terms soil composition and quality, with the context of its potentially permanent degradation from climate change and intensive land use, we will explore the independence of soils, and the ways in which they sustain life.

IMPRINTED MEMORY

There is a long history of the belief and understanding of soil as a life-sustaining element, from the first perceptions of the soil as a surface level film, enriched with organic substances, to the modern definition of profile and horizons. The material that constitutes soil is real, in the sense that it can be touched and felt. Defining soil is however very complex and has evolved throughout history.

The first documented relationships with soils were in the Mediterranean area by the civilizations developed in Egypt, the Middle East, India, and China around 7000-6000 BC. Soil was examined from an agricultural point of view and humans studied it merely to know which types of soils were more or less suitable for cultivation and production. For example, people discovered sandy soils were easy to prepare, but the production of crops was harder to maintain, while clay soils were harder to prepare, but the production was easier. Around 5000 BC in Egypt we can find the neolithic site of Tell el-Samara offering a testimony of such discoveries and evidencing new ways of planting and irrigating soils.

With the advent of Hellenic culture, the study and perception of soils embraced a philosophical or conceptual approach. Firstly, Herodotus (485-425 BC) considered it an important element to characterize a place. Subsequently, Aristotle (384-322 BC) divided the world created from amorphous matter into four elements: fire, air, water, and earth. The earth, which for him was nothing more than the soil, is defined as characterized by opposite qualities such as hot and cold, dry and humid, heavy and light, and hard and soft. Theophrastus (371-286 BC) instead distinguished for the first time *edaphos* (the earth intended as soil) from the Earth as a planet. He defined the soil in several levels: the surface, rich in organic substance, the subsoil, which is a source of nutrients for the roots, the substrate which is a source of fluids for plant roots, and the tartar, a dark and deep chasm.

2 - Visual timeline. (Image courtesy of NAHR)

During the period of the Roman Republic (509-27 BC) through to the time of the Roman Empire (27 BC-476 AD), the cultural vision takes a different turn, tending to the classification and schematization of all knowledge. Catone il Censore (234-149 BC), within his ***De Agriculture***, paid great attention to plowing and fertilization for the development of plants, classifying arable soils with nine main groups divided into twenty-one secondary ones. Varro il Reatino (116-27 BC) subsequently invented a classification of as many as 300 types of soils using properties that are still used in the field of pedology, and also school classification such as humidity, texture, presence of stones, color, and compaction. Columella (4-70 AD) wrote twelve volumes on agriculture. Within them, he established a new and further classification of soils based on the combination of some of his properties, which, however, he does not consider exhaustive. He was one of the pioneers of field experiments, aimed at addressing the problem of the loss of soil fertility due to overexploitation.

After the Middle Ages, the Renaissance, and the Enlightenment, the 19th-century theories regarding soils are numerous and based on different developed scientific disciplines. The geologists highlighted the relationships between rock and soil; the geomorphologists the relationships between the various landscape morphologies and the soil; the botanists those between vegetation and soil, and so on.

It was the work of Russian geologist and geographer, Vasily Vasilievich Dokuchaev (1846-1903), which established the foundations for modern soil science. Between 1873 and 1875, severe drought devastated the Russian prairies with very serious socio-economic repercussions. Under Tsar Alexander II, Dokuchaev, who was the geologist of the University of St. Petersburg, took charge of carrying out field studies in the area. From these investigations, he published fourteen books including the famous *Russian Chernozem* in 1883, and the new and current 'genetic' school of soil science was born. Here emerged the grounding principle of soil as an independent body (with specific morphologies, properties, and origins). Pedology, therefore, studies soils exactly as they are in their natural geographic context, in order to understand the regularity of their genesis, the interrelationships with soil-forming factors, and their geographical distribution. The Western concept of soil was changed radically and it was no longer just an object of agricultural activity. The knowledge of the soil as both a substance and a support for life were developed only after this radical change in viewpoint. Dokuchaev considered the soil as a natural body which holds its own importance. Not just because we need it. Soil carries its own dignity as information of the past keeper. Since soil's specific morphologies and properties are related to its own origins, soil is a keeper of information about the past conditions which determined its morphologies and properties.

SOIL AS SUBSTANCE

A first simple observation can allow us to describe soil as the material present on the surface of the Earth. Over time many definitions have been expressed, starting with Dokuchaev's definition (1883). Soil has been described as the growth medium of plants (Lyon & Buckman 1922), as a mantle of loose and weathered rocks (Shaler 1891), as the stimulated skin of the sub-aerial portion of the earth's crust (Nikiforoff 1959), as a system in spontaneous movement towards a state of equilibrium (Chesworth 1973), and as a reactor, memory, and regulator of the interactions of the biosphere (Targulian & Sokolova 1996). More recently, soil has been described as a component of the biosphere with ecological functions responsible for biodiversity and productivity (Dobrovolskii et al. 2001). The soil has also been described from a more thermodynamic point of view as a four-dimensional structure (including time and space) in which exist the residues and products of the alteration of the lithosphere, the dead and living organic substance of the biosphere and finally the elements coming from the atmosphere (Boulaine 1975). The organization and evolution of its structure are the results of forms of energy that manifest themselves on the surface of the globe.

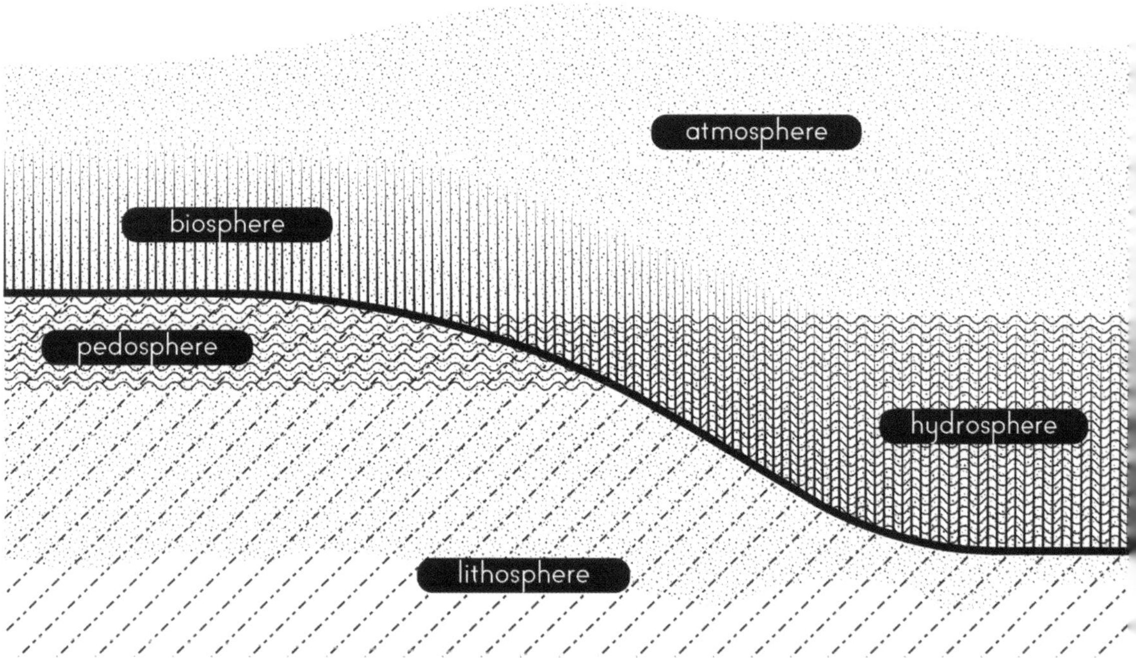

3 - Section view of atmosphere, hydrosphere, pedosphere, biosphere, and lithosphere. (Image courtesy of NAHR)

The soil is, therefore, a chemical-physical system which is open and complex, whilst also part of another complex system that exchanges matter and energy with what surrounds it in a continuous evolution (Boulaine 1975). Cumulatively this imprinted historical information led me to develop this definition of the soil as a three-dimensional independent geological body that is at the interface between the atmosphere, hydrosphere, biosphere, and lithosphere. This definition entails an independence of the discipline of soil studies from others. Moreover, even if studied independently, by interfacing and connecting with these four other spheres, it both influences and is dependent on their changing states.

The atmosphere is a gaseous shell that surrounds a celestial body, whose molecules are held in place by the body's gravitational force; it interacts with soils exchanging gasses and providing life for all living beings in soils. By hydrosphere we mean all kinds of water present in the various states of aggregation on our planet, from the subsoil to the surface up to the layers of the atmosphere: it interacts with soil providing elements from the atmosphere and weathering minerals in soil such as sand which helps the proliferation of life. The biosphere represents the set of places between the bottom of the ocean to the top of the first ten kilometers of atmosphere where life exists. It interacts with the soil, exchanging elements and providing carbon that otherwise would not complete its cycle. Finally, the lithosphere is the stiffest outer part of planet Earth composed mainly by rocks and deposits. It is the basis on which soil develops and provides the elements contained in soil from the beginning.

Even if from a scientific perspective soil is a 3D body, we can say that it covers the terrestrial surface like a film, with the exception of rocky outcrops, barren areas such as deserts, or those covered with ice. We can say this because the thickness, the third dimension, is so small compared to others, and to the Earth's radius, that it can be considered irrelevant. Soil originates as a result of the transformation of rocks that are found in a chemical-physical environment different from that of origin, whose transformation takes place in ways that are controlled by environmental factors and provides all elements and nutrients to sustain life. This original material (whether it is the geological substrate or a deposit) is called parent material.

SUSTAINING LIFE

The mother rock, not altered and which remains in-depth, can be termed parent material, as mentioned. The first band in which the rock begins to be transformed and can also be seen with the naked eye is called saprolite. Poorly weathered saprolite often contains aquifers that are capable of producing groundwater, often suitable for livestock. Yields depend on the texture of the materials and their depth from which the aquifer is derived. The product of the transformations happening in the saprolite is what we have defined as soil. The combination of soil and saprolite has its own name and is called regolith. The presence of regolith is one of the important factors for the survival of life, since few plants can grow on or within the solid rock, and animals would not be able to dig or build shelter without loose material.

During the formation of soil, two different moments can be distinguished, that of meteorization and that of pedoplasmation. The first involves the destruction, transformation, and all those processes that lead from a parent material to a saprolite, thus going from an organized system to a heterogeneous one. The second, on the other hand, involves the transition from saprolite to layered horizons (layers parallel to the surface which have different chemical-physical compositions, types, and characteristics) and the differentiation of these horizons themselves, thus returning from a disorganized to a heterogeneous system. The processes that are related to this differentiation, called pedogenetic processes, depend mainly on five factors: climate, biological forms, parent material, topography, and time. These factors leave important footprints within the soil that can be traced in the various components.

4 - Sectional diagram of soil, saprolite, regolith, and parent material. (Image courtesy of NAHR)

Generally, soil is made up of 45% minerals, 5% of organic matter, 25% water, and 25% air; these percentages can vary over time and depend on all the factors that lead to the formation of the soil (Dazzi 2021). The mineral fraction can be divided into primary or inherited constituents and secondary (newly formed) constituents. The first are those minerals of larger size that were present in the mother rock and that have been preserved in the soil. The second ones are instead those smaller minerals that were not present in the mother rock but were formed in the soil as a result of the transformations of the primary minerals. They are also called alteration complexes. The organic fraction, on the other hand, is composed of fresh organic matter as an origin, which can be vegetable residues, animal remains, or excrement.

Water is essential for the life of plants and all organisms present in the soil. It is not pure, but it contains dissolved hundreds and hundreds of organic and inorganic substances which constitute the so-called Soil Solution. We can say that if the soil is the skin of our planet, the solution of the soil is its blood! The movement of the solution is effectively one of the engines of pedogenesis and the movement or retention is limited by the size of the pores.

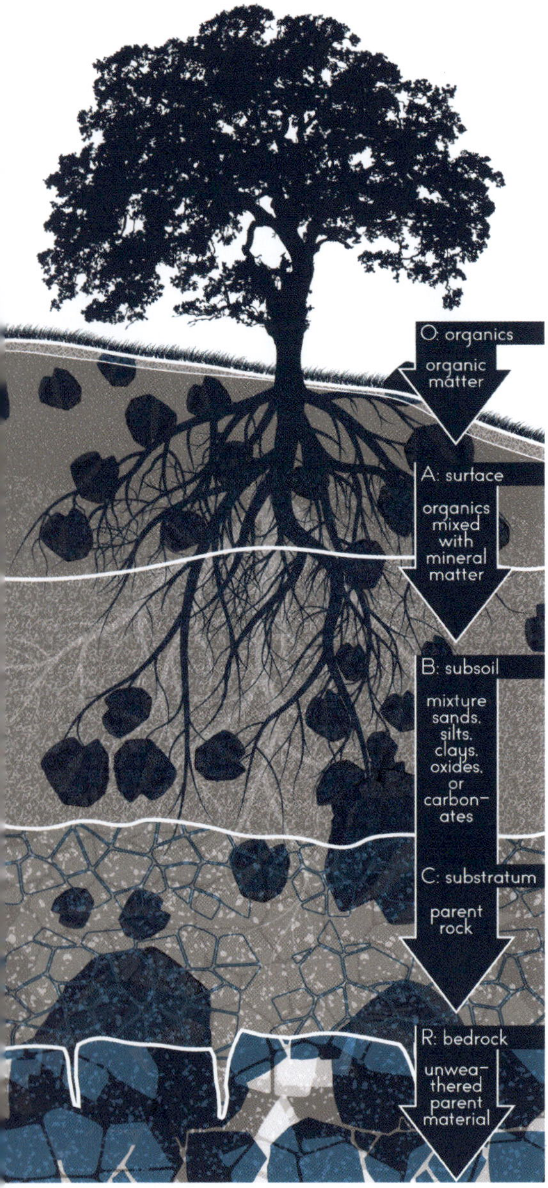

5 - Ideal soil profile and horizon sequence, with respective principal components. (Image courtesy of NAHR)

The air in the soil, on the other hand, occupies all those pores that are not occupied by water, so its quantity depends on particle size, porosity, and water content. Its composition depends on exchanges with the external atmosphere, the respiration of organisms, and nitrogen fixation. The air undergoes seasonal variations and influences the oxidation-reduction potential, and therefore the behavior of the chemical elements. In this way, the air makes some ions available (or not) for all the inhabitants of the soil, playing a fundamental role in life.

Changing the content in water and air may be dangerous as they are both essential to life and they occupy the same space; the pores. Too much water will result in too little air and vice versa, so they should be in balance, as they are naturally. Changes in organic matter content in undisturbed soils are very slow and usually take centuries to manifest. The cultivation of land, which implies the partial or total removal of the vegetation cover and the introduction of tillage, strongly influences the balance of organic matter, causing a rapid lowering of its original level. This can be countered by adopting appropriate agronomic practices aimed at enhancing the flow of carbon to the soil like the use of forage with grazing rather than cutting, the lengthening of the vegetative stages of the crops, the restoration of grassland and/or pasture crops, and the maintenance of the soil at colder temperatures through the covering action of the plants. If good practices are not adopted, we can incur what is known as land consumption.

Land consumption is a term used to describe the conversion of healthy soil into areas of industrial agriculture, road building, and human settlement. The European Environment Agency distinguished two different land-consuming activities that can be described as the expansion of built-up areas, and the land exploitation by agriculture, forestry, and other economic activities (Marquard 2020). Land loss is often used instead of land consumption, but formally it should only refer to the conversion of coastal land to open water caused mainly by coastal erosion, inadequate sediment supply, subsidence, and sea level rise.

The soil is also a fraction of the 'Critical Zone'. Critical Zone here refers to the volume of the interface between the top of the vegetation and the deep limit of the groundwater that supports life on Earth. It is therefore nothing more than a mixture of air, water, living organisms, organic matter, and geological materials. It is a system of chemical, biological, physical, and geological processes that work together to support life on the Earth's surface. Although our understanding has increased over the past hundred years, multidisciplinary and multi-scale approaches are still needed in order to integrate the understanding of its processes. The current problem of the Critical Zone derives from the fact that the nourishment of the human population requires tens of millions of square kilometers of cultivation. Today, 0.23 hectares of arable land is present per capita but with the current rate of land consumption and natural erosion over the next 40 years, only 20% of the necessary arable land will be available (not counting other problems such as desertification). This is because the alteration has a higher rate than the formation of saprolite while of course, the two processes should counterbalance each other.

During the last decades, soil properties and the overlaying vegetation have changed drastically due to the well-known effects of climate change.[1] Both abiotic and biotic components of the environment are showing ample evidence that climate change has affected a broad range of organisms with diverse geographical distributions.

6 - Infestation of *Robinia pseudoacacia*, an invasive species, in an Italian forest, 2022. (Image courtesy of author)

[1] In recent years a general awareness of the fact that the Earth's climate is changing was born, and, already in 1988, the establishment of an international committee of experts to assess the relevance, the speed, and possible response strategies have been founded by the General Assembly of the United Nations, called the IPCC (International Panel on Climate Change). In this context of climate change, the Alpine environment is particularly frail, and effects of climate change are evident both on biotic and abiotic factors. Among the abiotic components of cold terrestrial ecosystems, the cryosphere, which is the dominant element of these environments, showed striking and evident responses to climate change. During the last decades, the worldwide glacier retreat and mass decrease have strongly increased above the long term averages, with the European Alps as one of the most sensitive areas, coupled with a reduction of the snow cover depth and its areal extension (Thompson, 2016), due to the increase in summer temperature and concomitant decrease in winter precipitation.

7 - Compacted urban soil: details of the scarce porosity and thickness of the surficial horizons, 2018. (Image courtesy of author)

According to quantitative estimates of the biological impacts of the climate change 'fingerprint', a greater impact is expected on ecosystems and vegetation at high latitudes and altitudes such as the ones of the Alpine range. The greater influence of abiotic, rather than biotic, factors on regulating the function of these ecosystems, is strongly constrained by low temperatures, short growing seasons and frequent freezing events during the snow-free period. The expansion of exotic species from parks and gardens is another potential threat posed by climate change and can already be observed in the Southern Alps where evergreen trees (even palm trees!) are invading the natural forests. The species composition of plant communities could change with consequences as yet unknown for the whole food chain and at the interaction between these abiotic and biotic factors stands the soil—the basis of life on our planet. These changes in the composition of plant communities with the expansion of exotic species threaten the composition of fungal communities of the mycorrhizae. All these changes affect the billions of microorganisms present in the soils (millions of organisms are living in a teaspoon of soil), changing nutrients' availability.

In the same time period, the development of urban centers and cities has affected soil properties and distribution. We are continuing to expand urban areas, often with a low population density, and irreversibly cover natural and agricultural areas with asphalt and concrete, buildings, roads and other infrastructures, commercial, productive and service settlements. It is incredible that in Italy, in 2020, 2,143,208.9 ha of land was consumed, 5,174.7 ha more than in 2019, and that this increase is constant from 2016 (Latini and Zanon, 2021). These transitions, in addition to determining the permanent and irreversible loss of fertile soil, in most cases, cause further negative impacts, such as fragmentation of the territory, reduction of biodiversity, alterations of the hydro-geological cycle, and microclimatic modifications.

Finally, intensive agriculture has affected soil in the last decades. The overexploitation of soil dramatically changes the carbon and water content of soils, making them unavailable for further use and requiring the use of chemical fertilizers that are often even more harmful to soils on a longer time scale. Moreover, the ecological consequences of intensive agriculture are extremely serious and these can include deforestation, desertification, the use of pesticides and fertilizers, pollution of groundwater, acid rain, and consequent climate change.

8 - Agricultural soils: very light coloured and poor inorganic matter, 2021. (Image courtesy of I. Mazzoleni)

CONCLUSION

Soil plays an extremely important role in historical memory. Past climatic events can be understood by observing and studying traces of frost or drying, floods from the richness in clay, up to the anthropogenic effects that can be represented by artifacts or high concentrations of chemicals into items where they shouldn't be. History serves us in many ways; scientifically the research conducted in the past allows us to keep evolving the understanding of soil, but also memory imprinted in the soil itself, can help us reflect on the memory and knowledge we have of the world around us. In this critical period of climate change, knowing how our Earth has reacted in the past is essential to understanding what we will be facing and will assist us to take timely actions on our consumption and use of soil, dictated principally by our lifestyle. We also know more about the complexity of soil and how it depends on and influences other spheres. We can't think about soil in isolation if we want to improve our wellness and that of the whole world.

While there is always more to learn and discover, we hold to the understanding that there is a huge amount of life underground, not only above ground, and this is not a mere fact, but a condition that will impact all lives. The loss of the not easily visible microcosms of the soil will mean losing the biological processes, creatures, material, and support that sustains all life.

In learning from science we can begin to see how critical the shift is from the utilitarian extractive mode of soil use, in particular through consumption or agriculture. Actions and pollutants that impact soil's biotic and abiotic factors have been reducing the mere availability of soil, and at the micro level have impoverished its biota, putting all living organisms at risk and threatening the soil's capacity for sustaining life.

We have learned through our own imprinted memory that knowledge is acquired progressively, over time. We have enough knowledge now to act upon what we know.

Voices of the Fellows
Soil

Roxanne Korda, Editor

Soil as a knowledge-keeper rejuvenates and energizes stories of country and culture: of home. Memories are buried within soil horizons, each layer a repository of ancient wisdom and knowledge. Land is where history and laws are kept. Soils differ regionally and are shaped culturally. How might we find a place of belonging through reciprocal relationships with soil?

Fellows-in-residence from vast regions across the world were invited to study, sample, feel, observe, and listen to the soil to bring awareness to the current state of soil and the relationships communities have with soil through their interpretations and professional input.

The reflections to follow offer glimpses into the insights of select Fellows through a series of dialogues formatted around the following points: On Soil Engagement, On Soil Communication, On Soil & the Body, On Soil as Substrate, and On Difference: Soil & Dirt.

These conversations offer to unpack soil characteristics and secrets through creative and scientific practices unearthing soil as the critical zone.

1 - Robin Nanney. *Mirror Box*, 2015. (Image courtesy of artist)

ON SOIL ENGAGEMENT

The ecology of soil structure, and its biomes are important not only for food production for humans, but for the health of the entire planet. We live on 'Planet Earth'—it says it all!
Petra Regent, Visual Artist

I am constantly amazed by how different soils are in terms of texture and color, and how this relates to the hidden substructure of the bones beneath the soil—the rocky layer that gave rise to the soils. Subconsciously, I can tell what county I am in by the soil color—the bright brick red soils of South Devon; the brown soils of Fife; and the dark red soils of South Wales.
Petra Regent, Visual Artist

Soil is a world teeming with living information. We can attempt to listen with our ears, our bodies, and our thoughts. We can feel the soil and notice how it changes with the seasons and the weather. We can look at the soil with our eyes and microscopes, finding the depth of scale and explosion of life. Recognition of this life beneath our feet enriches our existence and brings us a far deeper understanding of the cycles of life and death as our bodies, food, water, and air are filtered through the soil.
Roxanne Korda, Opera Singer & Researcher

2 - Rave-inspired flyers for a post-anthropocene plant and soil disco. Isabel Kuh. *Discoterra*, 2022. (Image courtesy of artist)

Bacteria, fungi, and yeasts make life possible in many of its aspects. To introduce (or to reintroduce) fermentation in contemporary artistic practices allows us to establish links and relational practices while generating an opening towards multi-species collaborative artistic creation and invites us to connect with the invisible microscopic worlds.
Marco Ranieri, Artist

root and microbe

root and soil aggregates

microbes, roots, soil aggregates

roots and microbe

3 - The edaphon, the biota of the soil, and studies into its potential to act as yeast for the fermentation of flour in breads. Process images showing roots, microbes, and soil aggregates; and final bread loaf. Marco Ranieri. *Edaphic Fermentations*, 2022. (Images courtesy of artist)

My practice with soil is shaped by deep listening. Listening to the land, by my roots in an agricultural county in southwestern Ontario, Canada and my ancestral roots in Quaker cooperative farmers and people who toiled in the dirt, in direct observations and lived experiences as a sensitive and curious person growing up in a polluted place and seeing the impacts of industrial mono-culture and agri-culture on the land and also how life regenerates to heal itself, in my friends, colleagues, collaborators, mentors, and teachers in whatever form they take.
Danielle Stevenson, Environmental Scientist

Muddy Visions

In a remote future
I won't be bothered by flies
And I won't be scared by wolves
Because I'll be myself a fly and I'll be myself a wolf.

In a remote future
Cows won't be just slaughter meat
and orchids will no longer be home decorations
because I'll be myself a cow and I'll be myself an orchid.

In a remote future
migrants will no longer be marginalized
and bacteria will no longer be invisible
because I'll be myself a migrant and I'll be myself a bacterium.

The thought placed in every inch of the body will break its boundaries...
The skin will no longer be a film separating the inside from the outside...
It will be porous soil that absorbs and gives,
land of vital pollutions,
fertile fur that warms and embraces.

4 - Performance and poem. Emma Dotti. *Muddy Visions*, 2022. (Image courtesy of I. Mazzoleni, poem courtesy of artist)

My practice is inspired by the intrinsic beauty of biological processes occurring at micro-scales. In my project *Soil Secrets*, I am exploring soil microbiome interconnections with surrounding ecosystem components. For example, processes at tiny scales between human-soil, plant-soil, root-soil, rock-soil...
Ñemikafe, Textile & Embroidery Artist

5 - Ñemikafe. *Soil Secrets*, 2022. (Images courtesy of artist)

ON SOIL COMMUNICATION

Pattern language is a set of solutions that fit inside each other—we can look at the way that relationships in the soil build community, cycle resources, and heal to form deeply connected resilient systems. I don't just want to learn about healthy soil communities—I want to learn from and with them, too. Microbes and soil dwellers exist in a mosaic of diversity depending on the conditions of their hyper-local habitat. They have learned over time to build powerful and enduring functional relationships throughout their community, in place. Perhaps being native to nowhere, through convening with the soil communities, I can be rooted and at home wherever I am.
Anne LaForti, Biomimicist

I engage with and share the stories in the soil mostly by helping people test their soil for heavy metals and other contaminants, showing a city or region's soil contamination levels in virtual interactive maps, and developing and sharing plain language resources about how to understand and what to do about soil contamination (such as through the Healing City Soils project).
Danielle Stevenson, Environmental Scientist

The qualities of walking, of pausing, represent the first founding act: transmission and subtle dialogue to build. What resonates below performs the landscape above. Inside things are animated moods, which communicate without excluding us. With these qualities one communicates in the manner of the invisible, with all realms.
Giuseppe Mongiello, Visual Artist

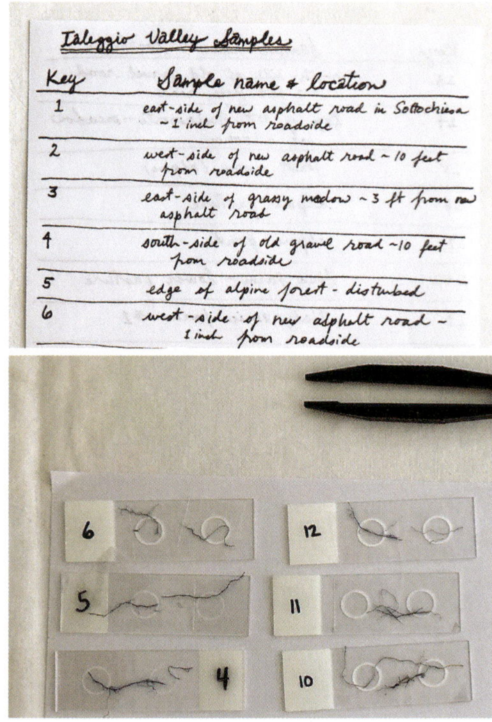

6 - Danielle Stevenson. *Seeing Fungi: Our Underground Benefactors*, 2022. (Image courtesy of scientist)

7 - Performing the landscape - A key from the soil organ keyboard. Clare Harmon. *Soil Organ*, 2022. (Image courtesy of artist)

Earth pigments are warm and subtle. When you try to make a usable pure pigment from the soil you really have to get to know it intimately. By grinding and sieving, you feel the silica, other minerals beneath the pestle. Often the pigment that emerges may not correspond to the original color of the soil.
Petra Regent, Visual Artist

8 - Painting using earth pigments and grass ink collected from the Taleggio Valley (A). Handmade print of grasses and yellow ochre pigment from the high pastures in the Taleggio Valley (B). Petra Regent. *Lombardy Landscape*, 2022. (Images courtesy of artist)

9 - *Coexistence* exhibition in Bergamo City, Italy, curated by NAHR. Roxanne Korda. *Listening to the Soil*, 2022. (Image courtesy of M. Macconi)

To detect the soil's language I buried my microphone. To create an exchange with the soil the signals were translated into musical motifs and sung back to the ground, incorporating movements which connected different parts of my body to the earth. The translated recordings search for a dramatic arc and trajectory, unique to the sounds of the soil, which can be explored through an expressive encounter with the earth, engaging literally and emotionally with it.
Roxanne Korda, Opera Singer & Researcher

ON SOIL & THE BODY

I am interested in the exploration and direct experience that people can have of the land in which they live, and the potential of art practice to promote narratives other than mainstream ones and that can increase ecological awareness, care for the land, more sustainable and balanced ways of living as both human and non-human.
Annalisa Zegna, Artist, Researcher & Cultural Practitioner

To consider, venerate, and act in communion with soil as an ancestor is to question where the border of the human body is located, and what purpose that border serves. To allow ourselves to be named as descendants of the earth's skin, is to shatter the colonial imposition of self-containment.
Anju-Marie Kasturiraj, Experimental Performance Director & Filmmaker

10 - Giuseppe Mongiello. *Mantica Exercises*, 2022. (Image courtesy of artist)

Steps, arms, feet, palms: the body cannot exceed the limits and measurements of its limbs, in their imprecision lies a vital breath. Popular Metrology and Meteorology are tools to resize our illusory hold on time and resources. Observation precedes action. Contemplation is the human's work.
Giuseppe Mongiello, Artist

Human eyes are above the ground but during the nomadic centuries before sedentarization we used to see with our feet as well, a form of knowledge we completely forgot and that in Western cultures has been confined to experiences such as hiking, detaching bodies from their primary domestic space.
Marco Ferrari, Architect

To uncover my soil secrets, I am using an empirical approach based on sampling on different soil ecosystems. Soil samples are dissected under the microscope to reveal marvelous forms, shapes, and colors for depicting interconnections with the microbiological world in an attempt to open a dialogue about the importance of soil microbiome for the entire terrestrial ecosystem.
Ñemikafe, Textile & Embroidery Artist

11 - Danielle Stevenson. *Seeing Fungi: Our Underground Benefactors*, 2022. (Image courtesy of scientist)

I enjoy zooming in and out many times over. When learning about the soil I get to see the intricate details and also how they fit together to impact our global food systems. Everyone eats, and therefore we're all connected to the soil and food systems in a deeply personal way—down to the microbes that live on and in our bodies.
Anne LaForti, Biomimicist

The texture, moisture, and temperature of soil are more intimately delivered through the feet so that it evokes the performer's imagination and performative power. The presence of soil itself emits an aura that is distinguished from other materials that serve as a floor. Touch is the most intimate and instinctual sensation that is often forgotten in language. I cannot directly implement the sensations I felt in the audience's body, however, I can make them go through similar experiences by setting up a visual and emotional journey, carefully and instinctually designed, that may lead to tactile experiences. An architect, Juhani Pallasmaa stated that "all the senses, including vision, can be regarded as extensions of the sense of touch—as specializations of the skin". Thus, visual and audible cues can evoke tactile sensations. In addition, mirror neurons allow us to share our experiences indirectly. I set my performances as a means to evoke tactile sensations in the audience's body using both physical display of the senses as well as visual narratives that touch emotional responses...
Oh-Kyung Jang, Performance Artist

12 - Oh-Kyung Jang. *Tactile Sensuality*, 2022. (Image courtesy of artist)

ON SOIL AS SUBSTRATE

Landscapes are as architectural as buildings, and we should start recognizing their inherent spatial content. This spatiality of the open field introduces a dimension based on porous boundaries and soft spaces' definition, introducing new stories and challenging concepts in contemporary architecture.
Marco Ferrari, Architect

Soil consumption is not an objective and steady/permanent concept. On the contrary, it evolves over time and space. It is especially affected by economic factors, and by small groups of privileged holders of power, in each area, place, country, and historical era. These groups want just to preserve their economic interests which, seldom, coincide with the public good. Who is considered a protector of the soil, or who is not, in a specific historical period could be different, if businesses and little groups' interests change.
Sonia Arienta, Dramaturgist & Writer

13 - Photo of the layout of a typical Italian apartment - the amount of soil lost in Italy every second. Sonia Arienta. *Public Puzzle - Act IV of Vuoti d'erba*, 2022. (Image courtesy of artist)

In Italy, we are losing 60% of national soil through public works. This is the outcome of the excessive presence of cars (the highest number in Europe) and the urban sprawl. Both are direct effects of the disregard and short-sightedness in the management of the public good.
Sonia Arienta, Dramaturgist & Writer

Our inability to see nature as a living organism and a life blinds us to recognize its right to exist and persist. It is beyond our rights to claim the ownership of nature, in which human beings are included, and its most basic components. The potential and the value of its usage can be calculated, nevertheless, we need another measure to perceive soil. This is something similar to what we call dignity of human beings. As a matter that makes up and provides the most basic component of our ecology, we need to start practicing to perceive soil as a living thing that deserves dignity. Once we develop this approach to connect to the soil, we can think outside of the boxes constructed by capitalism and anthropocentrism that assign a right to the human to dominate nature. Then, education becomes the area where we can imagine how to treat soil other than property and sustain a hope to develop different relationships with soils.
Oh-Kyung Jang, Performance Artist

My mother passed down traditions and responsibilities from our agricultural community in Tamil Nadu [India], which include understanding soil as a living ancestor. Right now, I am devising a South Indian ponkal ceremony to take place outdoors in England. Ponkal ritual includes boiling milk until it spills over in a piece of earthenware, usually to welcome a new home. I am making the earthenware out of soil from London and conducting the ceremony in the tradition of necro-ecology. I believe this reflexive interrogation of soil as a memory-keeper is essential if we are to heal the land in England, as it is home to the migrants, the refugees, and the descendants of the formerly enslaved, as it is a vessel for the masters and soldiers.
Anju-Marie Kasturiraj, Experimental Performance Director & Filmmaker

14 - Anju-Marie Kasturiraj. *Le Madri*, 2022. (Image still courtesy of artist)

ON DIFFERENCE: SOIL & DIRT

Soil is a living ecosystem. Dirt could become soil in the presence of a healthy microbiome.
Ñemikafe, Textile & Embroidery Artist

The difference between dirt and soil lies in their opposite operational dimension, in the different agencies they possess. Dirt is usually conceived as what is left, a passive element following a process of ruination or waste, while soil is what generates, as an active force of construction: actually, the major building site on Earth.
Marco Ferrari, Architect & Artist

Soil and dirt are defined by spatial relations, filth is in the eye of the subject and on the body of the 'other'.
Anju-Marie Kasturiraj, Experimental Performance Director & Filmmaker

Soil—as an active mixture of minerals, organic matter, air, water, and source of energy—supports all life, which in turn creates the messy mistakes, the feared and distrusted component—dirt—which ultimately will be restored to soil.
Roxanne Korda, Opera Singer & Researcher

Dirt gets dirty, sucks in the inorganic, difficult to listen to: transforms the fertile into a sterile landscape, a slow process in search of trust, recognition, and osmosis between breaths; complexity in transformation for its own sake. Soil is an immediate beauty and mutual feeling.
Giuseppe Mongiello, Artist

Soil is a big container of life, dirt is just a part of it, and in a linguistic way—in English but not in Italian language—it is much nearer to a devalued object, because of its assonance/association with the meaning of dust, garbage, powder.
Sonia Arienta, Dramaturgist & Writer

15 - Ritual. Julia Tröscher. *In the Dusk of Future Life*, 2022. (Image still courtesy of artist)

I believe that each language brings a specificity that is not easily translatable into different languages. For me, the term soil has a broader and deeper connotation than dirt: soil is a living ecosystem in continuous transformation, it is a vital process and not just materiality.
Annalisa Zegna, Artist, Researcher, & Cultural Practitioner

In my perspective, dirt is something related to the anthropic, with the refuse/discards of human activity. The soil isn't dirty. Soil is a living matter. In which it concentrates a big portion of the plant's life and biodiversity.
Marco Ranieri, Artist

Dirt has a different meaning to different people, but I think dirt is typically considered dead, just some broken down rocks, whereas soil is the living skin of the earth—a complex ecosystem with many interconnected parts including that weathered rock (sand, silt, or clay) and organic matter, air, water, and a whole bunch of living beings from worms to fungi, and bacteria and more to which human lives are inextricably intertwined.
Danielle Stevenson, Environmental Scientist

Groundings

Ilaria Mazzoleni

La Val Taleggio e' un'isola di pace in un mare di verde.

—Gianfranco Mazzoleni (circa 1978)

I believe we all yearn for a place that offers a deep sense of belonging in the world. For me, that place is the Val Taleggio. I was born in the village of Sottochiesa: "under the church." My first encounters of the world were green mountains and mountain houses designed by my father and managed by my mother. And this place provided a grounding for my own self: a place from which to begin exploring the world, and a place where I continue to return to, a home.

Belonging to home allows one to go anywhere, empowered to explore other places. I can embrace the discomfort of being elsewhere because I know there is comfort somewhere in the world: at home, here. The grounding of home allows for the freedom to leave. Going away is not a problem because coming back is always possible. The privilege of a two-way road is not available to everyone. I am privileged to explore our biodiverse world and when I return, it is to the land where I belong: the sky, mountains, woods, fields, riverbeds, animals that make up the special place of my valley. And this place has been a provocation for sharing.

The act of sharing is the act of sustainability, for me. It is through sharing not only our physical resources, like a place to be, but also of our intellectual capital, through open discourse which enriches our lives. Meeting someone that asks you to play with the handmade ocarinas in a public space challenges the notion of the self; to become the collective us, a super-organism that takes on a different behavior and collectively shapes the context in which it operates. For me, this togetherness is fundamental to sustainability and an attempt to put oneself aside in trying to comprehend the other and the wholeness that group is trying to embody. This effort might be a momentary action; it is not critical that there be permanent outcomes, for the shift still happens and remains within the individuals, who are capable of taking changes to the next place.

Through the leaving and returning journeys, I have also learned to become adaptable and flexible; to be responsive to global trends and note how these might locally impact lives in the Valley. At home, there is a constant reminder to appreciate, for example, the slowness of the geology of mountain peaks and valleys. There are many speeds; mostly all indifferent to us, as humans. The significance instead belongs in the tempo, and as with music, the tempo is critical even when asynchronous.

I am continuously learning from the Prealps topography. As a practicing architect, I feel the responsibility to respond to buildings' unresolved ecological tensions. Since the start of my career, I have felt the urgency of architecture to change as it contributes so much to carbon emissions. My interests in ecology, life cycle assessment of architectural materials and processes, bio-inspired design, and sustainable planning have driven my interest in working with nature: to adjust my pace to the climate pace.

I am learning to respect, deeply, the other; not only living, but all elements. Disturbance is something we as humans, seem great at generating, while attuning ourselves to contexts, especially of the environment. The silence of accommodations is a strength of other elements of nature. As I walk in the woods, I am fascinated by the majesty with which trees coexist, exchanging nutrients and communicating, by finding their ecological niche and participating in the greater life of the woods. I am enchanted every time the wind gently, or strongly, makes their branches swing to accommodate its passage and carrying away of seeds, pollens, and other organisms elsewhere. The biodiversity of nature is more than simple enchantment; it is the perfection of beauty and the inspiration to design, create, perform, by taking cues from it, trying to fit within it, lessening disturbance while being active participants with an agent of gentle change. The creativity inspired by the mountain landscapes and all their inhabitants is something I desire sharing with others.

The apparent simplicity by which things in nature merely occur, without drama or excess, simply because concurrent elements generate the happening of the next step, is magical. We can traverse the complexity of the world unfolding before us with such magic and a little alchemy, if we are able to act gently and proactively for purposes of conservation and care through the sharing of sustainable practices.

Portrait of the Valleys

I'm learning this place with my feet,
my map of this valley sketched with the rhythm of old knees.
I pack my bag for the day with water, fruit and ten words of Italian
then follow my curiosity as it winds uphill, a meadow pilgrimage
accompanied only by questions and this stingless sun.

—*Rachael Mead*, 'Pacing myself'

A snapshot of the NAH*R* locations reveals certain commonalities. Each Valley is rich in biodiversity, and invites inquiry-driven investigations through contemplations of mind, body, and soul. These are places and spaces to rest, research, work, discover, consider, and most significantly, to generate changes within, and with others, through unexpected encounters with natural spaces and cultures of staggering beauty.

1 - Taleggio Valley from Monte Zucco, 2017. (Image courtesy of I. Mazzoleni)

Taleggio Valley, Bergamo, Italy

Situated in the Province of Bergamo, two hours north of Milano, the Taleggio Valley lies in the Italian Orobie Prealps. Its dramatic scenery and secluded physical environment is a rich source of inspiration for bio-inspired projects.

The Valley's landscape has distinct and varied features at different elevations, from the bottom of ravines to the mountain pastures. A large section of the Valley includes the mountain reserve, *Parco delle Orobie*, which constitutes an important hotspot for biodiversity and nature conservation from high mountain air to heavily covered forest lowland and valleys (European Environment Agency n.d.). Recent climate events are affecting the animals and also human activities in the region. For example, droughts and loss of streams have led to more frequent wildfires, and the impact on various animals' habitats is yet to be documented.

The villages and towns offer many opportunities to observe and study the environments of both natural phenomena and human-made construction specific to the material culture developed over centuries. Recent impacts of climate change,

2 - Village of Sottochiesa, Taleggio Valley, Bergamo, Italy. NAHR headquarters, 2015. (Image courtesy of I. Mazzoleni)

agricultural patterns, land use change, forest management, lifestyle shifts, energy sources, technological opportunities, as well as legends and traditions about this territory, are just a few of the many themes that promise to inspire complex reasoning, critical evaluation, design responses, and other creative investigations.

The Soggiorno

Sottochiesa is one of the Valley's villages and in its center is NAHR's host and home for reflection, rest, and creative work: the *Soggiorno Mazzoleni*. Its main building was constructed in 1912 and acted as a hotel for over 50 years, then was refurbished in the late 1960s and turned into small apartments. Their simple but nostalgic charm has been a source of inspiration and background to the Fellows as they explore the surrounding landscapes. NAHR's activities are often hosted by the *Ecomuseum Val Taleggio*. This museum is a partner institution whose mission is to preserve the nature and the culture of the area, to promote a programme of ecologically aware events, and to explore a sustainable way of life in the mountains for future generations.

3 - The Soggiorno Mazzoleni (L) and garden (M), NAHR Home, 2022. (Image courtesy of I. Mazzoleni) Baita and Breakfast of Ecomuseum Val Taleggio (R), 2017. (Image courtesy Val Brembana Web)

Santa Ynez Valley, California, USA

The Santa Ynez Valley is about thirty miles north of Santa Barbara in an inland valley, with rolling Californian mountains rimming heavily cultivated farmland. The Santa Ynez Mountains are on the south, the San Rafael Mountains on the north, with the Santa Ynez River running from east to west where it flows into the Pacific Ocean.

From a geological point of view, the consolidated rocks of the Tertiary Period compose the surrounding hills. The Santa Barbara coast has a mediterranean climate and a complex topography. It regularly succumbs to wildfire as it is prone to downslope windstorms. Ecologically, the river provides a rich habitat for various endangered birds. The vegetation consists mainly of brush interspersed with chaparral, live oak, and grassland.

The farmland is mostly planted with wine grapes and the Valley is known for its wineries, as the soil in the Valley is well suited to the cultivation of certain grapes. The agriculture also includes significant groves of olive trees and fruit trees. The farms are typically small to medium-sized. There is also a notable community of horse owners and equine-serving businesses. The Valley attracts visitors who come to experience a typical California viticulture landscape, taste and purchase wine, visit the small towns, and enjoy the temperate climate.

4 - Santa Ynez Valley, Northwesterly view from Los Padres National Forest, 2015. (Image courtesy of Niranjan Arminius)

The Ranch

The residency's setting, *Refugio Road Ranch* or *r3*, is a ten-acre property, used for boarding retired horses, hosted by the distinguished architects Marc Angélil and Sarah Graham of *agps architecture*, Zurich and Los Angeles. The space is a refurbished farm structure that offers approximately 1,750 square feet of habitation, work, and exhibition space. Referred to as the *99¢ Space*, it is an airy and majestic farm shed structure, ideal for choreographic projects, projection installations, sculpture, or other visions that require ample space.

5 - Aerial view of the r3 ranch. NAHR-*Ca studio outpost*, 2022. (Image courtesy of E. Staudenmaier)

6 - The *99¢ Space* at r3, NAHR-*Ca studio outpost*, 2017. (Image courtesy of S. Graham and M. Angélil)

References

Journeying *NAHR*'s Transect
(p. xiv)

- Braidotti, R 2017, 'Posthuman, All Too Human: The Memoirs and Aspirations of a Posthumanist', paper presented at the The Tanner Lectures on Human Value, Yale University, 1-2 March, viewed 15 November 2023, <https://tannerlectures.utah.edu/_resources/documents/a-to-z/b/Braidotti%20Lecture.pdf>.
- Deleuze, G & Guattari, F 1993, *A Thousand Plateaus*, Minneapolis: University of Minnesota Press.
- Haraway, DJ 2003, *The companion species manifesto : dogs, people, and significant otherness*, Prickly Paradigm, Chicago, IL.
- Haraway, DJ 2016, *Staying with the Trouble: Making Kin in the Chthulucene*, Duke University Press, Durham, NC.
- Tsing, AL 2021, *The mushroom at the end of the world: On the possibility of life in capitalist ruins*, Princeton University Press, Princeton NJ.
- Wilson, EO 1984, *Biophilia*, Harvard University Press, London.

Bibliography

- Barrett, E & Bolt, B 2007, *Practice as research: approaches to creative arts enquiry*, Bloomsbury Visual Arts, London.
- Benyus, JM 2002, *Biomimicry: innovation inspired by nature*, Perennial, New York, NY.
- Morton, T 2007, *Ecology without nature: rethinking environmental aesthetics*, Harvard University Press, London & Cambridge, MA.
- Teti, V 2022, *La restanza*, Giulio Einaudi editore, Torino.

Ch1 - WOODS

Introduction *(p. 3)*

- Leonardi, C & Stagi, F 2019, *The Architecture of the Trees*, Princeton Architectural Press, New Jersey.
- Leonardi, C & Stagi, F 1982, *L'architettura degli alberi*, Mazzotta, Milan.

Trees Need People and People Need Trees *(p. 6)*

- Accelerate Resilience L.A. 2023, *MLK Trees Zoom Orbit*, online video, YouTube, viewed 10 January 2023, <https://youtu.be/ITgWYBgbKvE>.
- Allsup, K 2020, *Extending the life of urban trees*, viewed 10 January 2023, <www.agrinews-pubs.com/2020/02/11/extebdubg-the-life-of-urban-trees>.
- Armstrong, K 2017, *I Feel Your Pain: The Neuroscience of Empathy*, viewed 10 January 2023 <https://www.psychologicalscience.org/observer/neuroscience-empathy>.
- Baker, RSB 2013, *Men Of The Trees*, HardPress Publishing.
- Coutts, AM, White, EC, Tapper, NJ, Beringer, J & Livesley, SJ 2016, 'Temperature and human thermal comfort effects of street trees across three contrasting street canyon environments', *Theoretical and applied climatology*, vol. 124, no. 1, pp. 55-68.
- Heaviside, C, Macintyre, H & Vardoulakis S 2017, 'The Urban Heat Island: Implications for Health in a Changing Environment', *Curr Environ Health Rep.*, vol. 4, no. 3, pp. 296-305, viewed 16 March 2023, <https://pubmed.ncbi.nlm.nih.gov/28695487/>.
- Jamei, E & Tapper, N 2019, 'WSUD and urban heat island effect mitigation', in Sharma, AK, Gardner, T & Begbie, D *Approaches to Water Sensitive Urban Design*, Woodhead Publishing, pp. 381-407.
- Katz, C 2017, *Bark beetles explode across warming continents; some forests may not recover*, viewed 16th March 2023 <https://treesource.org/news/climate/bark-beetles-climate-change/>.
- Kravčík, M, Pokorny, J, Kohutiar, J, Kováč, M & Tóth, E 2007, *Water for the Recovery of the Climate: A New Water Paradigm*, Krupa Print, Zilina.
- Lipkis, A & Lipkis, K 1990, *The Simple Act of Planting A Tree: A Citizen Foresters' Guide to Healing Your Neighborhood, Your

City, and Your World, Jeremy P. Tarcher, Inc, LA, viewed 5 December 2023, <https://www.treepeople.org/wp-content/uploads/2021/06/the-simple-act-of-planting-a-tree.pdf>.
- Lungman, T, Cirach, M, Barboza, EP, Khomenko, S, Msselot, P, Quijal-Zamorano, M, Nueller, N, Gasparrini, A, Urquiza, J, Heroes, M, Thondoo, M & Nieuwenhuijsen, M 2023, 'Cooling cities through urban green infrastructure: a health impact assessment of European cities', *The Lancet*, vol. 401, no. 10376, pp. 577-589, viewed 16 March 2023, <https://www.thelancet.com/journals/lancet/article/PIIS0140-6736(22)02585-5/fulltext>.
- Salabert, S 2022 *How Los Angeles Is Leading the Urban Tree-Planting Revolution*, viewed 10 January 2023, <https://www.outsideonline.com/outdoor-adventure/environment/urban-tree-planting-los-angeles/> .
- TreePeople 2012, *Miracle on Elmer Avenue - updated,* online video, YouTube, viewed 10 January 2023, <https://www.youtube.com/watch?v=bwcK8IWawY0>
- TreePeople 2016, *Transferring Lessons From Australia's Millennium Drought to California: Accelerating Adaption to Drought, Flood & Heat*, viewed on 24 April 2023 <https://www.treepeople.org/resource/transferring-lessons-from-australias-millennium-drought-to-california/>.
- Wade, N 2009, *We May Be Born With An Urge To Help*, viewed 10 January 2023, <https://www.nytimes.com/2009/12/01/science/01human.html>.
- Water Stories 2022, *Reviving Rivers, with Dr Rajendra Singh the Waterman of India - Trailer,* online video, viewed 11 April 2023, <https://youtu.be/2PMsiXkAmFk>.

The Woods and Us: Between Resilience and Interdependence *(p. 19)*

Bibliography

- Blasi, C, Burrascano, S, Maturani, A, & Sabatini, FM (eds.) 2010, *Ancient forests in Italy*, Ministerio dell'Ambiente e della Tutela del Territorio e del Mare, Rome, viewed 22 September 2023, <https://www.mase.gov.it/sites/default/files/archivio/biblioteca/protezione_natura/foreste_vetuste_it.pdf>.
- Bonato, L, Lasen, C, Minelli, A, Solari, M, Tomaselli, M, & Uliana, M 2007, *Mountain conifer woods (Habitat notebooks, n. 18)*, Ministero dell'Ambiente e della Sicurezza - Museo Friulano di Storia Naturale, Udine, viewed 22 September 2023, <https://www.mase.gov.it/quaderni-habitat-n-18-i-boschi-montani-di-conifere-un-mantello-di-sottili-aghi-verdi>.
- Bracco, F, Marchiori, S, Mason, F, & Zanetti, A 2001, *The forests of the Po Valley (Habitat notebooks, n. 3)*, Ministero dell'Ambiente e della Sicurezza - Museo Friulano di Storia Naturale,Udine,= viewed 22 September 2023, <https://www.mase.gov.it/quaderni-habitat-n-3-le-foreste-della-pianura-padana-un-labirinto-dissolto>.
- Giupponi, C, Galassi, S, & Pettenella, D 2009, *Definition of the method for the classification and quantification of ecosystem services in Italy,* Ministerio dell'Ambiente e della Tutela del Territorio e del Mare, Rome, viewed 22 September 2023, <https://www.mase.gov.it/sites/default/files/archivio/biblioteca/dpn_tavolo7_servizi_ecosistemici.pdf>.
- Küster, H 2019, *History of the woods*, Bollati Boringhieri.
- Lorenzi, M, Mangili, L, Galizzi, F & Mangili, F 2004, *Natura Bergamasca*, Provincia di Bergamo, Ferrari Editrice.
- Pagani, E & Scaramellini, G (eds.) 1993, *Economic and social history of Bergamo, The original characteristics of the Bergamo area,* Fondazione Legler per la storia economica e sociale di Bergamo, viewed 22 September 2023, <https://fondazionestoriaeconomicabergamo.it/pubblicazioni/storia-economica-sociale-bergamo/volume-1>.
- Provincia di Bergamo, 2013, *Forest management plans (PIF) of the Province of Bergamo, 2008-2015*, Provincia di Bergamo, viewed 22 September 2023, <https://www.provincia.bergamo.it/

cnvpbgrm/zf/index.php/servizi-aggiuntivi/index/index/idtesto/697>.
› Sessi, A, Pesce, E, Tagliabue, E & Banfi, E 1981, *Natura in Lombardia - Vegetazione*, Regione Lombardia.

Voices of the Fellows: **Woods**
(p. 33)

› Benjamin, W 1978, 'The Author as Producer', in P Demetz (ed.), *Reflections: Essays, Aphorisms, Autobiographical Writings*, Schocken Books, pp. 220-238.
› Disney, D 2021, 'Co-authoring Communitas: resistance as counter-valence in John Kinsella's shared texts', *Angelaki: Journal of the Theoretical Humanities*, vol. 26, no. 2, pp. 69-80.
› Esposito, R 2013, Community, Immunity, Biopolitics', Translated by Z. Hanafi, *Angelaki*, vol. 18, no. 3, 2013, pp. 83-90.
› -- 2011,. *Immunitas: The Protection and Negation of Life*, Translated by Z. Hanafi, Polit

---

Ch2 - **ROCKS & STONE**
Introduction *(p. 47)*

› Bjornerud, M. 2018, *Timefulness, How Thinking Like a Geologist Can Save the World*, Princeton University Press, USA.

The Picturesque Versus the Planetary
(p. 49)

› California Office of Environmental Health Hazard Assessment 2023, *CalEnviroScreen 4.0 Maps*, viewed 23 February 2023, <https://oehha.ca.gov/calenviroscreen/maps-data>.
› Clark T 2021, *Cement: how its carbon impact can be reduced' Construction News*, viewed 23 February 2023, <https://www.constructionnews.co.uk/sustainability/carbon-cementing-net-zero-22-11-2021/>.
› Hutton, J 2019, *Reciprocal Landscapes. Stories of Material Movements*, Routledge, Andover.
› Kazys, V 2009 *The Infrastructural City: Networked Ecologies in Los Angeles*, Los Angeles Forum for Architecture and Urban Design, Network Architecture Lab Graduate School of Architecture Planning and Preservation Columbia University, Barcelona.
› Kiel, M 2017, *Empire State & Building*, Actar, New York.
› Kiel, M 2021, *Unless: The Seagram Building Construction Ecology*, Actar, New York.
› Long, KR, Alpers, CN, Orlando, J & Orris, GJ 2021, *Update of the Mineral Resources Data System for California including Mineral Deposit Types*: U.S. Geological Survey data release, viewed 24 July 2023, https://doi.org/10.5066/P9HTERGK.
› Lumley, L 2019, 'The Invisible Bituminous Desert.' *Log* N47 (20191001), 24-30: pp. 24–30.
› Pratt, S 2016, 'Benchmarks: March 17, 1944: The most recent eruption of Mount Vesuvius', *Earth: The Science Behind the Headlines*, March 15, viewed 23 February 2023, <https://www.earthmagazine.org/article/benchmarks-march-17-1944-most-recent-eruption-mount-vesuvius>.
› Regione Lombardia & Provincia di Bergamo, *Proposta di Piano Cave In attuazione della Sentenza* TAR BS n. 611/2013, viewed 24 July 2023, https://www.provincia.bergamo.it/cnvpbgrm/images/Att_Estrattive/Nuovo_PC/Relazione%20Tecnica%20e%20relativi%20allegati%20-%20C434.pdf.
› Stone News 2019, 'Italy's natural stone products exports in 2018', *Stone News*, viewed 23 February 2023, <https://stonenews.eu/italys-natural-stone-products-exports-2018/> and
› —2020, *Italy's imports of raw marble and granite*, viewed 23 February 2023, <https://stonenews.eu/italy-s-imports-of-raw-marble-and-granite/>.
› —2022, *Arabescato Orobico | Marble*, viewed 5 May 2023, <https://www.thestoneindustry.com/materials/arabescato-orobico>.
› Unicalce-Azienda 2021, *Unicalce Azienda*, viewed 23 February 2023, <https://www.unicalce.it/azienda/>.
› Yusoff, K 2022, *A Billion Black Anthropocenes or None*,University of Minnesota Press, Minneapolis MN.

What Rocks Remember *(pp. 64)*

- Berra, F & Jadoul, F 1996, 'Norian serpulid and microbial bioconstructions: Implication for the platform evolution in the Lombardy Basin (Southern Alps, Italy)', *Facies*, vol. 35, pp. 143-162, https://doi.org/10.1007/BF02536961
- Bjornerud, M 2018, *Timefulness: How Thinking Like a Geologist Can Help Save the World*, Princeton University Press, NJ.
- Camera di Commercio di Bergamo n.d.,2020, *Pietra Di Berbenno*, viewed on 8 May 2023, <https://www.bg.camcom.it/pietre-originali-della-bergamasca/pietre/pietra-di-berbenno>.
- Cita, MB & Gelati, R 1990, *Alpi e Prealpi Lombarde Vol. 1*, BE-MA, Dipartimento di scienze della terra dell'Università di Milano, Italy.
- Forcella, F, Jadoul, F, Bini, A & Ferliga, C 2000, *Carta Geologica (Geological Map) della Provincia di Bergamo e Note illustrative*, Servizio territorio della Provincia di Bergamo, Dipartimento di Scienze della Terra dell'Università degli Studi di Milano, Centro di Studio per la Geodinamica Alpina e Quaternaria del CNR, viewed 8 May 2023 <www.provincia.bergamo.it/cnvpbgrm/zf/index.php/servizi-aggiuntivi/index/index/idtesto/2259>.
- Gaetani, M, Sciunnach, D, Bini, A & Rossi, S 2012, *Carta Geologica d'Italia - Foglio 076 - Lecco e Note illustrative*, ISPRA, Servizio Geologico d'Italia, Lombardia.

Ch3 - **WATER**

Troubled Waters: Looking Forward with an Eye in the Rearview Mirror *(p. 89)*

- Bandaranayake, S 1993, 'Sigiriya - City, Palace and Royal Gardens,' in *The Cultural Triangle of Sri Lanka*, Paris: UNESCO, pp. 112-135.
- Carson, R 1962, *Silent Spring*, Houghton Mifflin, Boston.
- Cressy, G 1958, 'Qanats, Karez and Foggaras', *Geographical Review*, vol. 48, no. 1, pp. 27-44.
- De Meulder, B & Shannon, K 2013, *Water Urbanisms East*, UFO3: Explorations in Urbanism,Park Books, Zurich.
- English, PW 1968, 'The Origin and Spread of Qanats in the Old World', *Proceedings of the American Philosophical Society*, vol. 112, no. 3, pp. 170-181.
- Institute for Economics and Peace 2020, *Ecological Threat Register*, viewed 1 May 2023, <https://ecologicalthreatregister.org/>.
- Gandy, M 2006, 'The bacteriological city and its discontents', *Historical Geography*, vol. 34, pp. 14-25.
- Guterres, A 2012, *Working Group 1 Report*, Intergovernmental Panel on Climate Change (IPCC), viewed 1 May 2023 <https://unric.org/en/guterres-the-ipcc-report-is-a-code-red-for-humanity/>.
- Inglis, JT (ed.) 1993, *Traditional Ecological Knowledge: Concepts and Cases*, Ottawa: International Program on Traditional Ecological Knowledge and International Development Research Center.
- Isaacman, A & Isaacman, B 2013, *Dams, Displacement and the Delusion of Development: Cahora Bassa and its legacies in Mozambique, 1965-2007*, University of Ohio Press, Athens, OH.
- Klein, N 2014, *This Changes Everything: Capitalism vs. the Climate*, Simon & Schuster, New York.
- Kolbert, E 2014, *The Sixth Extinction: An Unnatural History*, Henry Holt and Company, New York.
- Macy, C 2009, *Dams*, W. W. Norton and Company, New York.
- Marot, S 2022, *Taking the Country's Side*, Poligrafa, Lisbon.
- Mathur, A & Da Cunha, D 2014, *Design in the Terrain of Water*, Applied Research & Design, San Francisco <https://www.mathurdacunha.com/>.
- Pottier, C 2008, 'The Angkor Territory. A Site for Khmer capitals,' in *Siem Reap/ Angkor Urban Observatory*, Architecture, Heritage, Development, Ipraus 1

- Rinne, K 2011, *The waters of Rome: aqueducts, fountains and the birth of the baroque city*, Yale University Press, New Haven
- Roy, A 2020 'The pandemic is a portal', *The Financial Times*, 4 April 2020, viewed 1 May 2023, <https://www.ft.com/content/10d8f5e8-74eb-11ea-95fe-fcd274e920ca?eType=EmailBlastContent&eId=ec801575-2b4b-4eb2-a9d8-db99f26da806>.
- Shannon, K, De Meulder, B, Gosseye, J & d'Auria, V (eds.) 2008, *Water Urbanisms* UFO1,Sun Publishers, Amsterdam
- Shannon, K & Manawadu, S 2007, 'Indigenous Landscape Urbanism: The Case of Sri Lanka's Tank System' in *JoLA (Journal of Landscape Architecture)*, Autumn 2007, pp. 30-41.
- Solnit, R 2010, *A Paradise Built in Hell: The Extraordinary Communities That Arise in Disaster*, Penguin Books, New York.
- Thomas, WL (ed.) 1956, *Man's Role in Changing the Face of the Earth*, University of Chicago Press, Chicago.
- Turpin, T 2008, *Dam*, Reaktion Books, London.
- UNICEF/ WHO 2019, *1 in 3 people do not have access to safe drinking water - UNICEF, WHO*, viewed on 1 May 2023, <https://www.who.int/news/item/18-06-2019-1-in-3-people-globally-do-not-have-access-to-safe-drinking-water-unicef-who>.

Changing Water Cycles in the Alps

(p. 105)

- Adrian, R, O'Reilly, CZ, Zagarese, H, Baines, SB, Hessen, DO, Keller, W & Winder, M 2009, 'Lakes as sentinels of climate change', *Limnology and oceanography*, vol. 54, no. 6 part 2, pp. 2283-2297.
- Auer, I, Böhm, R, Jurkovic, A, Lipa, W, Orlik, A, Potzmann, R & Nieplova, E 2007, 'HISTALP—historical instrumental climatological surface time series of the Greater Alpine Region', *International Journal of Climatology: A Journal of the Royal Meteorological Society*, vol. 27, no. 1, pp. 17-46.
- D'Agata, C, Diolaiuti, G, Maragno, D, Smiraglia, C & Pelfini, M 2020, 'Climate change effects on landscape and environment in glacierized Alpine areas: retreating glaciers and enlarging forelands in the Bernina group (Italy) in the period 1954–2007', *Geology, ecology, and landscapes*, vol. 4, no. 1, pp. 71-86.
- Diolaiuti, G & Smiraglia, C 2010, 'Changing glaciers in a changing climate: how vanishing geomorphosites have been driving deep changes in mountain landscapes and environments', *Géomorphologie: relief, processus, environment*, vol. 1, no. 2, pp. 131-152.
- Fenoglio, S, Bo, T, Cucco, M, Mercalli, L & Malacarne, G 2010, 'Effects of global climate change on freshwater biota: A review with special emphasis on the Italian situation', *Italian Journal of Zoology*, vol. 77, no. 4, pp. 374-383.
- Labaree, LW (ed.) 1961, *The Papers of Benjamin Franklin, vol. 3, January 1, 1745, through June 30, 1750,* New Haven: Yale University Press, pp. 60–67
- Google Earth 2022, *Autorità di Bacino distrettuale del fiume Po, The Po River with its Tributaries*, viewed 17 October 2023, <https://webgis.adbpo.it/>.
- Guglielmin, M, Donatelli, M, Semplice, M & Serra Capizzano, S 2018, 'Ground surface temperature reconstruction for the last 500 years obtained from permafrost temperatures observed in the SHARE STELVIO Borehole, Italian Alps', *Climate of the Past*, vol. 14, no. 6, pp. 709-724.
- Hilmi, N, Ali, E, Carnicer Cols, J, Cramer, W, Georgopoulou, E, Le Cozannet, G & Tirado, C 2022, 'IPCC AR6 WGII cross-chapter paper 4: Mediterranean region', in *EGU General Assembly Conference Abstracts*, pp. EGU22-10590.
- Hoekstra, A 2011, *The water footprint assessment manual: Setting the global standard*, Routledge.
- Jenkins, M 2003, 'Prospects for biodiversity', *Science*, vol. 302, no. 5648, pp. 1175-1177.
- Manca, M, Torretta, B, Comoli, P, Amsinck, SL & Jeppesen, E 2007, 'Major changes in trophic dynamics in large, deep sub—

alpine Lake Maggiore from 1940s to 2002: a high resolution comparative palaeo—neolimnological study', *Freshwater Biology*, vol. 52, no. 11, pp. 2256-2269.
- Smiraglia, C, Azzoni, RS, D'Agata, CARLO, Maragno, D, Fugazza, D & Diolaiuti, GA 2015, 'The evolution of the Italian glaciers from the previous database to the New Italian Inventory. Preliminary considerations and results', *Geografia Fisica e Dinamica Quaternaria*, vol. 38, no. 1, pp. 79-87.
- Trezzini, F, Giannella, G & Guida, T 2013, 'Landslide and flood: economic and social impacts in Italy', *Landslide Science and Practice: Volume 7: Social and Economic Impact and Policies*, pp. 171-176.

Ch4 - **GRASSES**
Grass, Food, and Climate in a Time of Uncertainty

(p. 140)

- Costa Jr., C, Wironen, M, Racette, K & Wollenberg, E 2021, *Global Warming Potential* (GWP*): understanding the implications for mitigating methane emissions in agriculture. CCAFS Info Note. Wageningen, the Netherlands: CGIAR Research Program on Climate Change*, Agriculture and Food Security (CCAFS).
- Giambartolomei, G, Forno, F & Sage, C 2021, 'How food policies emerge: The pivotal role of policy entrepreneurs as brokers and bridges of people and ideas', *Food Policy*, Elsevier, doi.org/10.1016/j.foodpol.2021.102038.
- Global Panel on Agriculture and Food Systems for Nutrition (GLOPAN) 2020, *Future Food Systems: For people, our planet, and prosperity*, GLOPAN, London.
- Sage, C 2022, 'A Research Agenda for Food Systems Introduction', in C Sage (ed.), *A Research Agenda for Food Systems*, Edward Elgar, Cheltenham.
- Sage, C, Antoni-Komar, I & Kropp, C 2021, 'Grassroots initiatives in food system transformation: The role of food movements in the second 'Great Transformation'', in Kropp, C, Antoni-Komar, I & Sage, C (eds.), *Food system transformations: Social movements, local economies, collaborative networks*, Routledge, Abingdon, UK.
- United Nations Department of Economic and Social Affairs, Population Division 2022, *World Population Prospects 2022: Summary of Results*, UN DESA/POP/2022/TR/NO, 3.
- Ritchie, H & Roser, M 2020, 'Environmental Impacts of Food Production', OurWorldInData.or, <https://ourworldindata.org/environmental-impacts-of-food>.

Lessons of the Fields and Pastures: Traditional Environmental Knowledge and Reciprocal Relationships with Nature in the Italian Prealps

(p. 153)

- Anderson, EN 1996, *Ecologies of the Heart: Emotion, Belief, and the Environment*, Oxford University Press, New York.
- Bird-David, N 1999, 'Animism' Revisited: Personhood, Environment, and Relational Epistemology', *Current Anthropology* 40.S1: S67-S91.
- Bulleitt, R 2007, *Hunters, Herders and Hamburgers. The Past and Future of Human Animal Relationships*, Columbia University Press, New York.
- Descola, P & Lloyd, J trans. 2013, *Beyond Nature and Culture*, University of Chicago Press, Chicago.
- Gibson, JW 2010, *A Reenchanted World: the Quest for a New Kinship with Nature*, Picador Press, New York.
- Grasseni, C 2017, *The Heritage Arena: Reinventing Cheese in the Italian Alps*, Berghan Books, Oxford.
- Harvey, G 2005, *Animism: Respecting the Living World*, Columbia University Press, New York.
- Hutton, R 2009, *Blood and Mistletoe: The History of the Druids in Britain*, Yale University Press, New Haven.
- -- 1999, *Drawing Down the Moon: A History of Modern Pagan Witchcraft*, Oxford University Press, Oxford.

- Kimmerer, RW 2013, *Braiding Sweetgrass: Indigenous Wisdom, Scientific Knowledge, and the Teachings of Plants*, Milkweed Editions, Minneapolis, MN.
- Kirner, K 2015, 'Pursuing the Salmon of Wisdom: the Sacred in Folk Botanical Knowledge Revival among Modern Druids', *Journal for the Study of Religion, Nature and Culture* 9/4, pp. 448-482.
- Lombardini, G 2002, *Costumi e proverbi valtellinesi*. reprint of 1926 edition, ristampa per l'Archivio del Centro di Studi Alpini di Isolaccia Valdidentro.
- Luhrmann, TM 2020, *How God Becomes Real: Kindling the Presence of Invisible Others*, Princeton University Press, Princeton.
- Magliocco, S 2004, *Witching Culture: Folklore and Neopaganism in America*, University of Pennsylvania Press, Philadelphia.
- Sciorra, J 2015, *Built With Faith: Italian American Imagination and Catholic Material Culture in New York City*, University of Tennessee Press,Knoxville, TN.
- Shir-Vertesh, D 2012, 'Flexible Personhood: Loving Animals as Family in Israel', *American Anthropologist* 114/3, pp. 420-432.
- Taylor, B 2010, *Dark Green Religion: Nature Spirituality and the Planetary Future*, University of California Press, Berkeley, CA.
- Tazzoli, TU 1935, *La contea di Bormio. Vol. III: Le tradizioni popolari. Raccolta di materiali per lo studio delle Alte Valli dell'Adda, Sondrio*. Arti grafiche valtellinesi.
- Viveiros de Castro, E 1999, 'Comments on 'Animism' Revisited: Personhood, Environment, and Relational Epistemology,' *Current Anthropology* 40.S1:S79.

Ch5 - ANIMALS

Introduction *(p. 183)*

- Ritvo, H 2007, 'on the animal turn', *Daedalus*, vol. 136, no. 4, pp. 118–122, https://doi.org/10.1162/daed.2007.136.4.118.

I Exist in Us *(p. 186)*

- Bangstad, S, Nilsen, T & Eliseeva, A 2019, Thoughts on the Planetary: An Interview with Achille Mbembe, *New Frame*, 6 September 2019, viewed 19 December 2022, <https://www.academia.edu/40268677/Thoughts_on_the_planetary_An_interview_with_Achille_Mbembe>.
- Blanchot, M 1989, *The Space of Literature*, trans. A Smock, University of Nebraska Press, Lincoln.
- Blaut, JM 1993, *The Colonizer's Model of the World: Geographical Diffusionism and Eurocentric History*, Guilford Press, New York.
- Descola, P 2013, *Beyond Nature and Culture*, trans. J Lloyd, University Of Chicago Press, Chicago.
- Despret, V 2016, *What Would Animals Say If We Asked the Right Questions?* University Of Minnesota Press, Minneapolis.
- Feral Atlas 2021, viewed 18 December 2022, <https://feralatlas.org>.
- Funke, C 2010, *Tintenherz*, Oetinger Taschenbuch, Hamburg.
- Harari, YN 2014, *Sapiens: A Brief History of Humankind*, Penguin Random House, London.
- Haraway, DJ 2007, *When Species Meet*, University of Minnesota Press.
- Heidegger, M 1982, *On the Way to Language*, 1st edn, HarperOne, San Francisco.
- Hume, D 2007, *A Treatise of Human Nature: Volume I*, Oxford University Press, New York.
- Kohn, E 2013, *How Forests Think*, University of California Press, London.
- Latour, B 2005, *Reassembling the Social: An Introduction to Actor-Network Theory*, Oxford University Press.
- Massey, DB 2005, *For Space*, SAGE, London.
- Simard, S 2016, *How Trees Talk to Each Other?*, TED Talk, viewed 16 December 2022, <https://www.ted.com/talks/suzanne_simard_how_trees_talk_to_each_other?referrer=playlist-the_secret_lives_of_plants&autoplay=true>.

- Sagan, D 2013, *The Human is More than Human: Interspecies Communities and the New 'Facts of Life'*, viewed 18 December 2022, <https://culanth.org/fieldsights/the-human-is-more-than-human-interspecies-communities-and-the-new-facts-of-life>.
- Timofeeva, O 2018, *The History of Animals: A Philosophy,* Bloomsbury Academic, London.
- Tsing, AL, Swanson, HA, Gan, E & Bubandt, N (eds.), 2017, *Arts of Living on a Damaged Planet*, University Of Minnesota Press, Minneapolis.

Bibliography

- https://www.geo.fu-berlin.de/en/geog/fachrichtungen/anthrogeog/zelf/MitarbeiterInnen/Ozan-Zeybek/index.html

Our Unique Mountain Homes
(p. 198)

- change.org 2021, *No More Lead for Hunting. Stop Intoxicating Raptors in the Alps!* 2023, viewed 20 February 2023, <https://www.change.org/p/provincia-di-sondrio-stop-al-piombo-nella-caccia-basta-rapaci-intossicati>.
- European Butterfly Monitoring Scheme - eBMS n.d., *Welcome,* viewed 25 April, 2023, <https://butterfly-monitoring.net/>.
- Groff, C, Angeli, F, Baggia, M, Bragalant, N, Pedrotti, L, Zanghellini, P & Zeni, M (eds.) 2022, *2021 Large Carnivores Report,* Provincia Autonoma di Trento, Wildlife Department, Large Carnivores Division, Trento.
- IdroLIFEife n.d. 2016, *Idro LIFE Project*, viewed 20 February 2023, <https://idrolife.eu/en/>.
- LIFEife Wolfalps EU n.d.2020, *Life Wolfalps EU: Co-ordinated Actions to Improve Wolf-Human Coexistence at the Alpine Population Level,* viewed 20 February 2023, <https://www.lifewolfalps.eu/en/>.
- The Alpine Convention n.d., *Framework Convention: Guiding principles for sustainable life in the Alps,* viewed 17 February 2023, <https://www.alpconv.org/en/home/convention/framework-convention/>.
- The Alpine Convention 2020, *Declaration of the XVI Alpine Conference on the Protection of Biodiversity and its Promotion at an International Level,* viewed 20 February 2023, <https://www.alpconv.org/fileadmin/user_upload/Topics/ACXVI_MountainBiodiversityDeclaration_en.pdf>.
- UC Berkeley Events 2015, *Mission of the National Parks and its Relevancy Today (Part 1)*, online video, Youtube, viewed 18 February 2023, <https://www.youtube.com/watch?v=o3ys5Us1dRs&t=410s>.

Ch6 - **SOILS**

Introduction *(p. 231)*

- Montanarella, L, Scholes, R, & Brainich, A (eds.) 2018, *Assessment report on land degradation and restoration*, IPBES, viewed 19 December, 2023, <https://www.ipbes.net/assessment-reports/ldr>.
- Ritchie, H & Roser, M 2019, 'Land Use', *Our World In Data*, September, viewed 21 May 2023, <https://ourworldindata.org/land-use>.

Mother Earth: Coming Home
(p. 233)

- AIATSIS The Australian Institute of Aboriginal and Torres Strait Islander Studies n.d., *The Mabo Case,* viewed 8 March 2023,
- <https://aiatsis.gov.au/sites/default/files/research_pub/overturning-the-doctrine-of-terra-nullius_0_3.pdf >.
- Anderson, S 2016, *'Murray Sinclair: Education is key to reconciliation'*, viewed 10 April 2023, <https://www.kairoscanada.org/murray-sinclair-education-key-reconciliation>.
- BanffEvents, 2016, *Indigenous Knowledge and Western Science: Dr. Leroy Little Bear Talk*, online video, YouTube, viewed 10 April, 2023, <https://www.youtube.com/watch?v=gJSJ28eEUjl>.
- Buhner, SH 2002, *The Lost Language of Plants: The Ecological Importance of Plant Medicines to Life on Earth,* Chelsea Green Publishing Company White River Junction, Vermont.

- Cajete, G 2000, *Native Science; Natural Laws of Interdependence*, Clear Light Publishing, Santa Fe, New Mexico.
- Ermine, W 'Cree Religious Ethos', *Indigenous Saskatchewan Encyclopedia*, University of Saskatchewan, viewed April 10, 2023 <https://teaching.usask.ca/indigenoussk/import/cree_religious_ethos.php>.
- *Law of Property Bill* [Lords] HC Deb 15 May 1922, vol. 154, cc.. 89-175 <https://hansard.parliament.uk/commons/1922-06-16/debates/019a7fbc-612a-4f17-b668-44aa3517ac2f/LawOfPropertyBillLords>.
- Skolimowski, H1990, 'For the Record: On the Origin of Eco-Philosophy', *The Trumpeter*, vol. 7, no. 1, viewed 10 April 2023, <https://trumpeter.athabascau.ca/index.php/trumpet/article/view/511/866>

Imprinted Memory - Sustaining Life *(p. 245)*

- Boulaine, J 1975, *Geographie des sols*, Puf, Paris.
- Chesworth, W 1973, 'The Residua System Of Chemical Weathering: A Model For The Chemical Breakdown Of Silicate Rocks At The Surface Of The Earth', *European Journal of Soil Science*, vol. 24, no.1, pp. 69-81.
- Dazzi, C 2021, *Fondamenti di Pedologia*, Le Penseur, Potenza.
- Dobrovolskii, GV, Nikitin, ED & Karpachevskii, LI 2001, 'New Approaches to the Concept of Soil Place in the Biosphere', *Eurasian Soil Sciences*, vol. 34, pp. S1-S5.
- Dokuchaev, VV 1883, *Russkii chernozem*, St Petersburg.
- Latini, L & Zanon, S 2021, *Soil As a Landscape: Nature, crossings and immersions, new topographies*, Fondazione Benetton Studi Ricerche Antiga, Treviso.
- Lyon, TL & Buckman, HO 1922, *The nature and properties of soils: a college text of edaphology*, Macmillan, London.
- Marquard, E, Bartke, S, Gifreu i Font, J, Humer, A, Jonkman, A, Jürgenson, E, Marot, N, Poelmans, L, Repe, B, Rybski, R, Schröter-Schlaack, C, Sobocká, J, Tophøj Sørensen, M, Vejchodská, E, Yiannakou, A & Bovet, J 2020, 'Land Consumption and Land Take: Enhancing Conceptual Clarity for Evaluating Spatial Governance in the EU Context', *Sustainability*, vol. 12, no. 19, pp. 8269.
- Nikiforoff, CC 1959, 'Reappraisal of the Soil: Pedogenesis consists of transactions in matter and energy between the soil and its surroundings', ***Science***, vol. 129, no. 3343, pp. 186-196.
- Shaler, NS 1891, *The origin and nature of soils*, Washington Government Printing Office, Washington.
- Targulian, V & Sokolova T 1996, 'Soil as a biotic/abiotic natural system : A reactor, memory, and regulator of biospheric interactions', *Eurasian Soil Science*, vol. 29, pp. 30-41.

APPENDIX
Portrait of the Valleys *(p. 273)*

- European Environment Agency n.d., The Natura 2000 protected areas network, viewed 31 January 2023, <https://www.eea.europa.eu/themes/biodiversity/natura-2000>.
- Mead, R 2019, 'Pacing myself', in chlorophyll & casein, Self published, p. 9.

Editors

Maristella Casciato
Rome, Italy
Los Angeles, USA

An architect and architectural historian. She has been Senior Curator, Head of Architectural Collections at the Getty Research Institute in Los Angeles since 2016, and has taught the history of twentieth-century architecture for almost forty years. In the course of the past decade, she has reconciled her interests in architecture with a more interdisciplinary approach to landscape and the natural environment. Maristella joined NAHR's Scientific Committee in 2019. Her research projects have been supported by a Fulbright Fellowship (1992), by a grant issued by the Institut National d'Histoire d'Art (Paris 2004), by the Mellon Foundation (2010), and by the Graham Foundation (2016). Book contribution: **editor of woods chapter**.

Lucinda Coleman
Perth, Australia

A choreographer, writer, teacher, and an adjunct researcher and lecturer at Edith Cowan University, Australia. Her PhD explored dance-making as a site for social justice arts engagement, and her research interests also include collaboration, collective practices, site-specific performance-making, and cross-cultural performance work. She was the founding Dance Maker of the Australian performing arts collective, *Remnant Dance*, 2010-2020, and recipient of NAHR collective residency with *Remnant Dance* 2018, and a solo residency 2020 when she also joined as Ambassador. Book contribution: **general editor and editor of voices**.

Sara Invernizzi
Bergamo, Italy

A museographer and writer living in the Orobie Prealps. Her research focuses on the complexity of Alpine culture, investigated with a multidisciplinary gaze. Invernizzi collaborates with the Consortium for the Protection of Strachìtunt Valtaleggio and has been a member of the scientific committee of the Ca' Martì Museum since 2020. Sara is an Ambassador of NAHR and coordinates several of the onsite activities and the relations with the local public governing agencies. Book contribution: **editor of grasses chapter**.

Roxanne Korda
Birmingham, UK

An opera singer, librettist and performance practice researcher based in Birmingham, UK. Her work explores presenting new opera, on esoteric subject matter, in unique spaces, to different audiences. Having previously studied Physics and Philosophy she completed her PhD in performance based research in 2023 and co-founded the company Infinite Opera in 2018. Roxanne has been a NAH*R* Ambassador since 2021. Book contribution: **general editor and editor of voices**.

Ilaria Mazzoleni
Taleggio, Italy
Los Angeles, USA

An architect, educator, author, and NAH*R* Founder. Her research focuses on sustainability and biomimicry, where innovation in architecture and design is inspired by the processes and functions of nature. Her built and conceptual work in the fields of sustainable architecture and biomimicry has been published internationally. Mazzoleni has been a faculty member at the Southern California Institute of Architecture (SCI-Arc), CalPoly Pomona, and CalArts in Los Angeles. Mazzoleni has authored and published the book *Architecture Follows Nature - Biomimetic Principles for Innovative Design* (CRC Press, 2013). Book contribution: **chief editor, *Journeying*, and *Groundings* author**.

Richard Molina
Los Angeles, USA

An educator and the design director of ANIMOL, whose interests align with social, civic, and ecological undertones embedded within architecture. Current investigations favor spatial nuances conceived by recombining geometry and compositional techniques to achieve diverse ranges of enclosure and openness, while encouraging interconnectedness.
Richard is a Lecturer in the Department of Architecture at Cal Poly Pomona and a Faculty in the School of Theater at CalArts. He is an active member of the Association for Women in Architecture and Design (AWA+D) and a NAH*R* Ambassador. book contribution**: editor of soil chapter and manuscript design**.

Gabi Scardi
Milan, Italy

An art critic, curator and writer, and NAHR President since 2019. She investigates art and the public sphere and contemporary collaborative methodologies. Scardi collaborated internationally with museums and institutions, curating exhibitions and public projects. Since 2011 she has been the artistic director of nctm e l'arte project and since 2020 co-director of Animot review. She contributes reviews and features to many publications and exhibition catalogs. She teaches modules on Contemporary Art and Public Art at Università Cattolica and IED. Book contribution: **editor of animals chapter**.

Aslı Suner
Istanbul, Turkey

An architect and an educator. Her passion for design is informed by a constant dialogue between her design practice and her academic role bridging interests like the ecologies of building, urban nature, and the relationality of design across time and space. Before founding Designcase, she worked at international architectural firms like Cannon Design and Foster + Partners. Suner teaches interdisciplinary design classes, architecture studios, sustainable building technologies, and biomimicry at Kadir Has University, Istanbul. Aslı has been a NAHR ambassador since 2020. Book contribution: **editor of rock chapter**.

Deborah Weintraub
Los Angeles, USA

The highest ranking architect for the City of Los Angeles, and the Chief Deputy City Engineer. Since 2019, Deborah has served as the NAHR Scientific Committee Chair. Her focus is sustainably designed public buildings and open spaces. She has worked on City buildings, bridges, parks, historic restorations, the Los Angeles River, and on repositioning large infrastructural City elements, such as decommissioned reservoirs and flood basins. Book contribution: **editor of water chapter**.

Contributors

Cover Design

April Greiman
Los Angeles, USA

A transmedia artist, designer, whose projects actively challenge interdisciplinary backgrounds, investigating the parallels & intersections between art, design, color, light and space. NAH_Fellow 2017

Foreword

Malena Martinez
Lima, Peru

Physician and researcher, passionate about exploring the complexity of the Peruvian territory. Together with her brother, chef Virgilio Martinez, Malena co-directs projects and the research for their interdisciplinary group, *Mater,* aiming to approach the megadiversity of their country from a holistic basis, while developing a unique knowledge source on ingredients, traditions, ecology, and culture for creating gastronomic concepts, such as *Central* (2023 winner of World 50 Best restaurants awards), *Kjolle, Mil* in Peru, and *Maz* (2 Michelin stars 2023) in Japan, all of which represent Peru in different and unique ways. Malena is also the collaborator and/or co-author of culinary publications: *Central* (Phaidon 2016), *Mater Catalogue* (Editalo 2019), *Latinamerican Cookbook* (Phaidon 2021), *Mil* (Catapulta 2022) and *Festin* (Children's cookbook, Pichoncito 2023).

Preface and Journeying

Enrico Bassi (NAHR scientific committee member) is a naturalist and ornithologist, from Bergamo, Italy.

David Chan (NAH_Fellow '19) is a designer and artist living in Los Angeles, California.

Emilio Fantin (NAH_Visitor '22) is an artist who lives and works in Bologna, Italy.

Marco Ferrari (NAH_Fellow '22) is an architect and artist, cofounder of AIDEL, from Italy.

Yu-He Lin (NAH_Fellow '15) is a performance artist from Taipei, Taiwan.

Remi Picó (NAH_Fellow '23) is an Italian scientific researcher and artist based in Boston, USA.

Michael Rotondi, architect

Mauro Sacchi (NAH_Fellow '15) is an Italian dance-theater practitioner based in Taipei, Taiwan.

Andria Zachariou (EMPACT - NAH_Fellow '23) is an artist based in Famagusta, Cyprus.

Woods

Andy Lipkis is Project Executive of Accelerate Resilience Los Angeles. He founded TreePeople at 18, and started planting trees, and rehabilitating smog and fire damaged forests. He pioneered Urban/Community Forestry and Urban Watershed Management movements. His honorary titles include *Forester,* and *Landscape Architect* conferred upon him by The Society of American Foresters and the American Society of Landscape Architects respectively.

Luca Mangili is President of the Flora Alpina Bergamasca (FAB) Association and has professionally dealt with the construction and maintenance of green areas in Italy. He collaborates with Italian scientific institutions and local bodies on study projects and protection of **the floristic** heritage of the Orobie Prealps. Additionally, he has contributed to publications on flora and natural environments, which have been supported and promoted by FAB.

NAHR Fellows 2016

Caroline Clerc is a Los Angeles-based artist working in photography.

Dan Disney is an Australian poet based in Seoul, South Korea.

Isabelle Duvivier is an American-Belgian architect, based in Venice Beach, California.

Elke Eichmann is an architect and photographer from Zurich.

İdil Kemaloğlu is an architect based in Paris and Istanbul, Turkey.

Mauriah Kraker is a collaborative performance maker from Milwaukee, USA.

Viviana González is a Colombian fine artist based in Switzerland.

Jenny Rodenhouse is a multimedia designer and artist, based in Los Angeles, USA.

Jay Yowell is an architect and educator based in Edmond, Oklahoma.

Other Contributors

Marco Macconi, photographer.

Katura Reynolds, illustrator.

Surya Tüchler (NAH_Fellow '21) is a performance artist living in Hamburg, Germany.

Sarah Valeri (NAH_Fellow '23) is an artist, educator, and art therapist in Brooklyn, New York, USA.

Thanos Zakopoulos, creative naturalist.

Rock & Stone

Joshua G. Stein is Professor of Architecture at Woodbury University and the founder of Radical Craft, a Los Angeles-based studio that advances an experimental art and design practice saturated in history, archaeology, and craft.

Stefania Cabassi is a freelance Geologist. Born in Milan, she has lived in mountain contexts since 1999 and from 2005 in the Bergamo Orobie Prealps. There she works on Geological Risk Assessment, Land Planning and Protection, on behalf of several public administrations.

NAHR Fellows 2017

Juan Barte is a graphic designer and photographer from Madrid, Spain.

Katy Bradford is a Structural Engineer based in Atlanta, GA, USA.

Simon Eales Eco-Poet & Researcher from Melbourne, Australia based in Buffalo, NY, USA.

Kim Kwangwoo is an artist living in South Korea.

Sofia Malone is a visual artist based in Stockholm, Sweden.

Giulia Pellegrini is an Italian artist based in Brussels, Belgium.

Greta Ruedisueli is a landscape architect based in New York City, NY, USA.

Ho Ting Wei is an animation filmmaker and multidisciplinary artist based in Taipei, Taiwan.

Visitors and Other Contributors

Lisa Agnelli, photographer.

Anne Elvey (NAH_Visitor '17) is a poet and editor based in Naarm/Melbourne, Australia.

Luca Foresti, photographer.

Alice Guarisco (NAH_Visitor '17) is an Italian architect based in New York city, USA.

Pietro Invernizzi (NAH_Visitor '17) is a stone master based in Val Imagna, Italy.

Cinzia Invernizzi, architect.

Frank Melendez (NAHR_Fellow '23) is an architectural designer based in New York City, USA.

Water

Bruno De Meulder and Kelly Shannon are professors

of urbanism at KU Leuven, Belgium. Their design research focuses on the evolving relation of landscape, particularly water and forests, and urbanism. They have worked extensively in Belgium and Vietnam, developing multi-scalar scenarios that respond to global warming.

Silvia Picone is a researcher of the direct and indirect effects of recent climate change on permafrost dynamics, with a particular focus on permafrost hydrology. Silvia graduated with honors in the master's degree of Environmental Sciences at the University of Insubria, Italy, in 2020. She is soon to complete a PhD in Polar Sciences at the Ca'Foscari University, Italy, in collaboration with the University of Insubria.

NAHR Fellows 2018

Eunice Andrada is an award-winning Filipina poet and lyricist based in Sydney, Australia.

Jill Anholt is a sculptor and artist living in Vancouver, Canada.

Ellen Avery has worked as a Remnant Dance artist and lives in Australia.

Luisa Brando is an artist and architect living in Madrid, Spain.

Katie Chown has worked as a Remnant Dance artist and lives in Australia.

Carley Culmo is a performing artist from Florida, USA.

Tiffany Henschel is an interdisciplinary designer and researcher from Portland, Oregon.

Samuel Maddox is an architect and academic focused on ecological design and rural contexts, based in Boston.

Giulia Mattera is an artist researcher working with live art, ecology, and human-nonhuman collaboration.

Skye Morét is a designer and scientist in the USA focused on bridging the gap between nature and design.

Stacy Passmore is an architect and co-founder of Superbloom in Denver, Colorado.

Rachel Peachey & Paul Mosig are artists based in Australia working on the lands of the Gundungurra and Darug people.

Ashe Rodrigues is a Filipino-American multidisciplinary artist based in South Florida, USA.

Matthew Wong is a landscape architect and designer based in San Francisco, USA.

Brian Yulo Ng is a water treatment business lead and Filipino/Singaporean artist and filmmaker.

Visitors and Other Contributors

Armando Bottelli, photographer.

Ardito Desio, photographer.

Stefano Perona, photographer.

Susanna Ravelli (NAHR Scient. Com.) is a cultural project manager based in Cremona and Milano, Italy.

Grasses

Colin Sage is an independent research scholar who works on the interconnections of food systems, planetary health and human wellbeing. He is the author of *Environment and Food* (2012); co-editor of five books; and editor of *A Research Agenda for Food Systems* (2022).

Sabina Magliocco, Ph.D. is Professor of Anthropology and Chair of the interdisciplinary Program in the Study of Religion at the University of British Columbia in Vancouver, Canada.

NAHR Fellows 2019

Katie De Bari is a performing artist and choreographer based inFlorida, USA.

Anaïs Hazo is an artist, researcher and birth worker who operates between France and the Netherlands.

Ming Lin is a writer-researcher from New York, USA.

Heidi Maribut is a visual artist from Copenhagen, Denmark.

Rachael Mead is a poet, short story writer and arts critic living in South Australia.

Siobhan Rigg is an interdisciplinary artist based in Washington, DC, USA.

Floris van der Kleij is an Interdisciplinary Designer based in Amsterdam, The Netherlands.

Visitors and Other Contributors

Elena Brebenel (NAH_Fellow '16) is a textile designer from Ploiesti, Romania.

Courtlin Byrd (NAH_Visitor '19) is a filmmaker and a PhD Candidate Media Studies at SUNY-Buffalo.

Lucrezia Costa (NAH_Fellow '23) is an Italian artist based in Milano, Italy.

Brent Cox (NAH_Visitor '19) is a poet and PhD Candidate in Poetics at SUNY-Buffalo, NY, USA.

Giuseppe De Santis, photographer.

Ariana Ferguson (NAH_Fellow '23), dancer.

Marilyn L. Geary (NAH_Fellow '23) is an oral historian, author and fiber artist based in the San Francisco Bay Area, California.

Jeffrey M Hancock (NAH_Fellow '23) is a dancer and researcher based in Chicago, USA.

Danica Liongson (NAH_Fellow '22) is a landscape architect from Indianapolis, USA.

Rowan McCloskey (NAH_Fellow '23), dancer.

A. Monti (NAH_Visitor '19), poet.

Beth Robertson (NAH_Fellow '23) is a musician and artist from Glasgow, UK.

Sophie Teitler (NAH_Fellow '23), dancer.

Heloise Tunstall-Behrens (NAH_Fellow '23) is a musician and composer from London/Cornwall, UK.

<u>**Animals**</u>

Dr Sezai Ozan Zeybek is a human geographer based at Technische Universität, Berlin. He has two published books: *Animals in the Recent History of Turkey* and *Talking Through Tales: A Storytelling Guide*.

Cristina Movalli is a biologist at Val Grande National Park, Italy (2002 onwards), whose work encompasses managing projects around nature conservation and scientific research, and promoting environmental education to visitors and local residents. Her work focuses on studying the relationship between, and coexistence of, humans and other living beings as well as best practices and strategies for conservation and sustainability of these extreme and fragile natural environments.

NAHR Fellows 2021

Camilla Alberti is a visual artist from Milan, Italy.

Tami Banh is an architect and landscape architect based in New York City, NY, USA.

Meri Gorni is an interdisciplinary artist who lives in Paderno Dugnano, Italy.

Susan Haris is a multispecies ethnographer from India.

Xinning Liu is a Chinese artist and animator based in the UK.

Patrick L. Jaimes is a photographer and visual artist from Mexico City.

Francesco Maluta is a visual artist living and working in Milan, Italy.

Laleh Motlagh is an Iranian artist and educator based in Chicago, USA.

Emma O'Mara is an architectural designer and artist living in San Francisco, California.

Jonathan Pando Ocón is a biogeography researcher from the USA.

Marta Romani and Karl Logge are visual, performance and installation artists from Italy.

Visitors and Other Contributors

Silvia Crocetti, photographer.

Cornelia Funke is a German children's fiction author and illustrator living in Volterra, Italy.

Gianni Gritti, photographer.

Petra Gruber is an architect from Vienna, Austria.

Fulvia Mendini (NAHR scientific committee member) is an artist based in Milan, Italy.

Alessandro Mendini (NAHR scientific committee founding member) was an architect and artist from Milan, Italy.

Mario Monfrini, photographer.

Paola Nalin, photographer.

Simona Nelli, photographer.

Fiorella Pesenti Campagnoni is a farmer and photographer from Sottochiesa, Taleggio Valley, Italy.

Nora Sweeney, (NAH_Fellow '22) is an artist and filmmaker from Los Angeles, USA.

Yulia Shtern (NAH_Fellow '19) is a Canadian Artist based in Verona, Italy.

Marco Tessaro, photographer.

Stefano Torriani is a naturalist illustrator based in San Pellegrino Terme, Bg, Italy.

Soil

Marilyn Poitras is a Michif/ Irish/Scottish woman born and raised on the prairies. She holds a Master of Laws (Harvard Law School) and is a lifelong student of Indigenous Laws with Elders from home and around the globe. She is a systems disrupter and designer.

Alessandro Longhi is a postdoctoral geopedologist and adjunct professor at Università degli Studi dell'Insubria, Italy. Longhi's main focus of study has been the reconstruction of deglaciation history in the Central Italian Alps since the last glacial maximum, and on the soils that currently reside there.

NAHR Fellows 2022

Sophia Aitken is an urban planner and artist based in Florida, USA.

Sonia Arienta is a playwright and dramaturge living in Milan, Italy.

Lily Garget is an artist based in Glasgow, UK.

Clare Harmon is a translator, interdisciplinary artist and researcher based in Minneapolis, Minnesota.

Oh-Kyung Jang is a multi-disciplinary artist from South Korea.

Anju-Marie Kasturiraj is a video artist based in London, UK.

Isabel Kuh is an artist and curator from Los Angeles, USA.

Anne LaForti is a biomimicist based in California, USA.

Giuseppe Mongiello is a visual artist from Italy born in Vallo della Lucania.

Ñemikafe is an embroidery artist and biologist, based in Bonn, Germany.

Marco Ranieri is an artist, activist, and independent researcher based in Spain.

Petra Regent is a visual artist from Bristol, UK.

Angus Rushin is an artist based in Glasgow, UK.

Danielle Stevenson is a multidisciplinary environmental scientist based in California, USA.

Julia Tröscher is a multidisciplinary artist based in Antwerp, Belgium.

Annalisa Zegna is an artist and researcher from Italy.

Visitors and Other Contributors

Emma Dotti (NAH_Workshop Abitanza '22) is a visual artist based in Milan, Italy.

Robin Nanney (NAH_Fellow '15) is an architect based in Los Angeles, USA.

Portrait of the Valleys

Sarah Graham and Marc Angélil, architects.

Eric Staudenmaier, photographer.

Niranjan Arminius, photographer.

Val Brembana Web, photography.

Index

D

E

F

N

O

P

Q

R

S

T

Transect of Coexistence
Inquiry into Nature, Art, and Habitat

by NAHR
Nature, Art & Habitat
info@nahr.it
www.nahr.it

Chief Editor
Ilaria Mazzoleni

Editors
Maristella Casciato
Lucinda Coleman
Sara Invernizzi
Roxanne Korda
Richard Molina
Gabi Scardi
Aslı Suner
Deborah Weintraub

Cover design and photograph by
April Greiman

Internal cover image by
Richard Molina

Published by
ListLab
info@listlab.eu
listlab.eu

LISTLAB

ListLab Editorial Director
Alessandro Martinelli
asma.meetarch (IG)

ListLab Art Director & Production
Blacklist Creative
blacklist-creative.com

Sales, Marketing & Distribution
distribution@listlab.eu
listlab.eu/distributori

BABEL INTERNATIONAL This book is part of the BABEL International book series

ISBN 9788832080940

Printed and bound
in the European Union, July 2024

About ListLab
ListLab was established in 2007. Since then, it has elaborated on the idea of an international editorial laboratory with a multidisciplinary approach to architecture, planning, arts, photography, and design. **List Group**, founded in 2021, aims at creating networks and promoting debates and cultural exchange but also organizes events from which new knowledge about architecture, cities, and landscapes can develop. Today, List Group is composed of **ListLab**, the publishing house, **Blacklist**, the graphic design studio, **Instaura**, the informational weblog, and **Us/Them/Yours**, a creative agency that aims at a multimedia approach to information.

Born as an academic publisher, **ListLab** controls the quality and originality of the publications issued to disseminate adequate and up-to-date knowledge. For this purpose, **ListLab** has an Editorial Board and an Editorial Director that supervise the choice and development of the publications. Also, **ListLab** has an international Scientific Commission, the members of which operate or manage peer-review assessment of the publications. In particular, **ListLab** adopts a single-anonymized peer-review process with the contribution of at least two different reviewers. This book passed a round of peer-review process and was consequently edited.

For more information about the Editorial Board and Scientific Commission, please visit: *listlab.eu/board*